Larry Bull, Ester Bernadó-Mansilla and John Holmes (Eds.)

Learning Classifier Systems in Data Mining

Studies in Computational Intelligence, Volume 125

Editor-in-chief
Prof. Janusz Kacprzyk
Systems Research Institute
Polish Academy of Sciences
ul. Newelska 6
01-447 Warsaw
Poland
E-mail: kacprzyk@ibspan.waw.pl

Further volumes of this series can be found on our homepage: springer.com

Vol. 103. Devendra K. Chaturvedi
Soft Computing Techniques and its Applications in Electrical Engineering, 2008
ISBN 978-3-540-77480-8

Vol. 104. Maria Virvou and Lakhmi C. Jain (Eds.)
Intelligent Interactive Systems in Knowledge-Based Environment, 2008
ISBN 978-3-540-77470-9

Vol. 105. Wolfgang Guenthner
Enhancing Cognitive Assistance Systems with Inertial Measurement Units, 2008
ISBN 978-3-540-76996-5

Vol. 106. Jacqueline Jarvis, Dennis Jarvis, Ralph Rönnquist and Lakhmi C. Jain (Eds.)
Holonic Execution: A BDI Approach, 2008
ISBN 978-3-540-77478-5

Vol. 107. Margarita Sordo, Sachin Vaidya and Lakhmi C. Jain (Eds.)
Advanced Computational Intelligence Paradigms in Healthcare - 3, 2008
ISBN 978-3-540-77661-1

Vol. 108. Vito Trianni
Evolutionary Swarm Robotics, 2008
ISBN 978-3-540-77611-6

Vol. 109. Panagiotis Chountas, Ilias Petrounias and Janusz Kacprzyk (Eds.)
Intelligent Techniques and Tools for Novel System Architectures, 2008
ISBN 978-3-540-77621-5

Vol. 110. Makoto Yokoo, Takayuki Ito, Minjie Zhang, Juhnyoung Lee and Tokuro Matsuo (Eds.)
Electronic Commerce, 2008
ISBN 978-3-540-77808-0

Vol. 111. David Elmakias (Ed.)
New Computational Methods in Power System Reliability, 2008
ISBN 978-3-540-77810-3

Vol. 112. Edgar N. Sanchez, Alma Y. Alanís and Alexander G. Loukianov
Discrete-Time High Order Neural Control: Trained with Kalman Filtering, 2008
ISBN 978-3-540-78288-9

Vol. 113. Gemma Bel-Enguix, M. Dolores Jimenez-Lopez and Carlos Martín-Vide (Eds.)
New Developments in Formal Languages and Applications, 2008
ISBN 978-3-540-78290-2

Vol. 114. Christian Blum, Maria José Blesa Aguilera, Andrea Roli and Michael Sampels (Eds.)
Hybrid Metaheuristics, 2008
ISBN 978-3-540-78294-0

Vol. 115. John Fulcher and Lakhmi C. Jain (Eds.)
Computational Intelligence: A Compendium, 2008
ISBN 978-3-540-78292-6

Vol. 116. Ying Liu, Aixin Sun, Han Tong Loh, Wen Feng Lu and Ee-Peng Lim (Eds.)
Advances of Computational Intelligence in Industrial Systems, 2008
ISBN 978-3-540-78296-4

Vol. 117. Da Ruan, Frank Hardeman and Klaas van der Meer (Eds.)
Intelligent Decision and Policy Making Support Systems, 2008
ISBN 978-3-540-78306-0

Vol. 118. Tsau Young Lin, Ying Xie, Anita Wasilewska and Churn-Jung Liau (Eds.)
Data Mining: Foundations and Practice, 2008
ISBN 978-3-540-78487-6

Vol. 119. Slawomir Wiak, Andrzej Krawczyk and Ivo Dolezel (Eds.)
Intelligent Computer Techniques in Applied Electromagnetics, 2008
ISBN 978-3-540-78489-0

Vol. 120. George A. Tsihrintzis and Lakhmi C. Jain (Eds.)
Multimedia Interactive Services in Intelligent Environments, 2008
ISBN 978-3-540-78491-3

Vol. 121. Nadia Nedjah, Leandro dos Santos Coelho and Luiza de Macedo Mourelle (Eds.)
Quantum Inspired Intelligent Systems, 2008
ISBN 978-3-540-78531-6

Vol. 122. Tomasz G. Smolinski, Mariofanna G. Milanova and Aboul-Ella Hassanien (Eds.)
Applications of Computational Intelligence in Biology, 2008
ISBN 978-3-540-78533-0

Vol. 123. Shuichi Iwata, Yukio Ohsawa, Shusaku Tsumoto, Ning Zhong, Yong Shi and Lorenzo Magnani (Eds.)
Communications and Discoveries from Multidisciplinary Data, 2008
ISBN 978-3-540-78732-7

Vol. 124. Ricardo Zavala Yoe (Ed.)
Modelling and Control of Dynamical Systems: Numerical Implementation in a Behavioral Framework, 2008
ISBN 978-3-540-78734-1

Vol. 125. Larry Bull, Bernadó-Mansilla Ester and John Holmes (Eds.)
Learning Classifier Systems in Data Mining, 2008
ISBN 978-3-540-78978-9

Larry Bull
Ester Bernadó-Mansilla
John Holmes
(Eds.)

Learning Classifier Systems in Data Mining

With 65 Figures and 53 Tables

Springer

Larry Bull
School of Computer Science
University of the West of England
Bristol BS16 1QY
U.K.
Larry.Bull@uwe.ac.uk

John Holmes
Centre for Clinical Epidemiology
and Biostatistics
University of Pennsylvania
Philadelphia, PA 19104
U.S.A.
jholmes@cceb.med.upenn.edu

Ester Bernadó-Mansilla
Enginyeria i Arquitectura La Salle
Universitat Ramon Llull
08022 Barcelona
Spain
esterb@salleURL.edu

ISBN 978-3-540-78978-9 e-ISBN 978-3-540-78979-6

Studies in Computational Intelligence ISSN 1860-949X

Library of Congress Control Number: 2008924365

Cover design: Deblik, Berlin, Germany

Printed on acid-free paper

9 8 7 6 5 4 3 2 1

springer.com

Contents

Learning Classifier Systems in Data Mining: An Introduction

Larry Bull[1], Ester Bernadó-Mansilla[2], and John Holmes[3]

[1] School of Computer Science, University of the West of England, Bristol BS16 1QY, UK, Larry.Bull@uwe.ac.uk
[2] Enginyeria i Arquitectura La Salle, Universitat Ramon Llull, 08022 Barcelona, Spain, esterb@salleURL.edu
[3] Centre for Clinical Epidemiology and Biostatistics, University of Pennsylvania, Philadelphia, PA 19104, USA, jholmes@cceb.med.upenn.edu

Summary. This chapter provides an introduction to Learning Classifier Systems before reviewing a number of historical uses in data mining. An overview of the rest of the volume is then presented.

1 Introduction

There is now widespread recognition that it is possible to extract previously unknown knowledge from datasets using machine learning techniques. Learning Classifier Systems (LCS) [28] are a machine learning technique which combines evolutionary computing, reinforcement learning, supervised learning or unsupervised learning, and heuristics to produce adaptive systems. They are rule-based systems, where the rules are usually in the traditional production system form of "IF state THEN action". An evolutionary algorithm and heuristics are used to search the space of possible rules, whilst a credit assignment algorithm is used to assign utility to existing rules, thereby guiding the search for better rules. The LCS formalism was introduced by John Holland [28] and based around his more well-known invention – the Genetic Algorithm (GA) [27]. A few years later, in collaboration with Judith Reitman, he presented the first implementation of an LCS [31]. Holland then revised the framework to define what would become the standard system [29,30]. However, Holland's full system was somewhat complex and practical experience found it difficult to realize the envisaged behaviour/performance (e.g., [54]) and interest waned. Some years later, Wilson presented the "zeroth-level" classifier system, ZCS [50] which "keeps much of Holland's original framework but simplifies it to increase understandability and performance" [50]. Wilson then introduced a form of LCS which altered the way in which rule fitness is calculated – XCS [51]. The following decade has seen resurgence in

L. Bull et al.: *Learning Classifier Systems in Data Mining: An Introduction*, Studies in Computational Intelligence (SCI) **125**, 1–15 (2008)
www.springerlink.com

the use of LCS as XCS in particular has been found able to solve a number of well-known problems optimally. Perhaps more importantly, XCS has also begun to be applied to a number of hard real-world problems such as data mining, simulation modeling, robotics, and adaptive control (see [6] for an overview) and where excellent performance has often been achieved. Further, given their rule-based nature, users are often able to learn about their problem domain through inspection of the produced solutions, this being particularly useful in data mining. Formal understanding of how such systems work has also increased in recent years (see [8] for an overview). The purpose of this volume is to bring together current work on the use of LCS for data mining since this is the area in which they have experienced the most growth in recent years with excellent performance in comparison to other techniques [e.g., 2].

The rest of this contribution is arranged as follows: Firstly, the main forms of LCS are described in some detail. A number of historical uses of LCS in data mining are then reviewed before an overview of the rest of the volume is presented.

2 Holland's LCS

Holland's Learning Classifier System [30] receives a binary encoded input from its environment, placed on an internal working memory space – the blackboard-like message list (Fig. 1). The system determines an appropriate response based on this input and performs the indicated action, usually altering the state of the environment. Desired behaviour is rewarded by providing

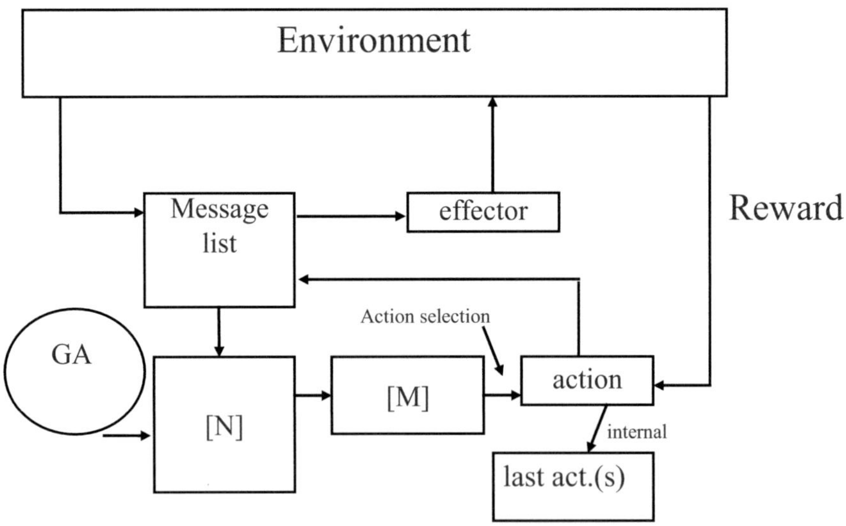

Fig. 1. Schematic of Holland's learning classifier system

a scalar reinforcement. Internally the system cycles through a sequence of performance, reinforcement and discovery on each discrete time-step.

The rule-base consists of a population of N condition-action rules or "classifiers". The rule condition and action are strings of characters from the ternary alphabet {0,1,#}. The # acts as a wildcard allowing generalisation such that the rule condition 1#1 matches both the input 111 and the input 101. The symbol also allows feature pass-through in the action such that, in responding to the input 101, the rule IF 1#1 THEN 0#0 would produce the action 000. Both components are initialised randomly. Also associated with each classifier is a fitness scalar to indicate the "usefulness" of a rule in receiving external reward. This differs from Holland's original implementation [31], where rule fitness was essentially based on the accuracy of its ability to predict external reward (after [42]).

On receipt of an input message, the rule-base is scanned and any rule whose condition matches the external message, or any others on the message list, at each position becomes a member of the current "match set" [M]. A rule is selected from those rules comprising [M], through a bidding mechanism, to become the system's external action. The message list is cleared and the action string is posted to it ready for the next cycle. A number of other rules can then be selected through bidding to fill any remaining spaces on the internal message list. This selection is performed by a simple stochastic roulette wheel scheme. Rules' bids consist of two components, their fitness and their specificity, that is the proportion of non-# bits they contain. Further, a constant (here termed β) of "considerably" less than one is factored in, i.e., for a rule C in [M] at time t:

$$\text{Bid }(C,t) = \beta \cdot \text{specificity }(C) \cdot \text{fitness }(C,t)$$

Reinforcement consists of redistributing bids made between subsequently chosen rules. The bid of each winner at each time-step is placed in a "bucket". A record is kept of the winners on the previous time step and they each receive an equal share of the contents of the current bucket; fitness is shared amongst activated rules. If a reward is received from the environment then this is paid to the winning rule which produced the last output. Holland draws an economic analogy for his "bucket-brigade" algorithm (BBA), suggesting each rule is much like the middleman in a commercial chain; fitness is seen as capital. The reader is referred to [46] for an introduction to reinforcement learning.

The LCS employs a steady-state Genetic Algorithm operating over the whole rule-set at each instance. After some number of time-steps the GA uses roulette wheel selection to determine two parent rules based on their fitness relative to the total fitness of the population:

$$\text{Probabilty\_Selection }(C,t) = \text{fitness }(C,t)/\Sigma \text{ fitnesses }(t)$$

The effect of this scheme is to bias reproduction towards those rules which appear to lead to higher reward from the environment. Copies are made of the chosen rules which are then subjected to two genetic operators: mutation and crossover. Mutation is applied probabilistically at a per-locus rate (e.g., 1/100) along the length of the rule and upon satisfaction the value at that locus is altered – typically, a locus becomes one of the other two possible values with equal probability. For example, if the above mentioned rule 1#1:0#0 experiences a mutation event on its last locus it could become 1#1:0#1 or 1#1:0##. Crossover begins by randomly choosing a position within the rules and then swaps them from that point to their end. For example, the two rules 000:000 and 111:111 which experience crossover at position two would become 001:111 and 110:000 respectively. The purpose of the genetic operators is to introduce new rules into the population based on known good rules with the aim of discovering better rules. The new rules then replace two existing rules, often chosen using roulette wheel selection based on the reciprocal of fitness. The reader is referred to [16] for a recent introduction to evolutionary computing.

It is important to note that the role of the GA in LCS is to create a cooperative set of rules which together solve the task. That is, unlike a traditional optimisation scenario, the search is not for a single fittest rule but a number of different types of rule which together give appropriate behaviour. The rule-base of an LCS has been described as an evolving ecology of rules – "each individual rule evolves in the context of the external environment and the other rules in the classifier system." [18] A number of other mechanisms were proposed by Holland but for the sake of clarity they are not described here (see [32] for an overview).

3 Wilson's ZCS

As noted above, Wilson introduced the simple ZCS to increase understandability and performance. In particular, Wilson removed the message list and rule bidding (Fig. 2) and did not allow wildcards in actions. He introduced the use of action sets rather than individual rules, such that rules with the same action are treated together for both action selection and reinforcement. That is, once [M] has been formed a rule is picked as the output based purely on its fitness. All members of [M] that propose the same action as the selected rule then form an action set [A]. An "implicit" bucket brigade [22] then redistributes payoff in the subsequent action set.

A fixed fraction – equivalent to Holland's bid constant – of the fitness of each member of [A] at each time-step is placed in a bucket. A record is kept of the previous action set $[A]_{-1}$ and if this is not empty then the members of this action set each receive an equal share of the contents of the current bucket, once this has been reduced by a pre-determined discount factor γ (a mechanism used in temporal difference learning to encourage solution brevity (e.g., [46])). If a reward is received from the environment then a fixed fraction of this value is distributed evenly amongst the members of [A] divided

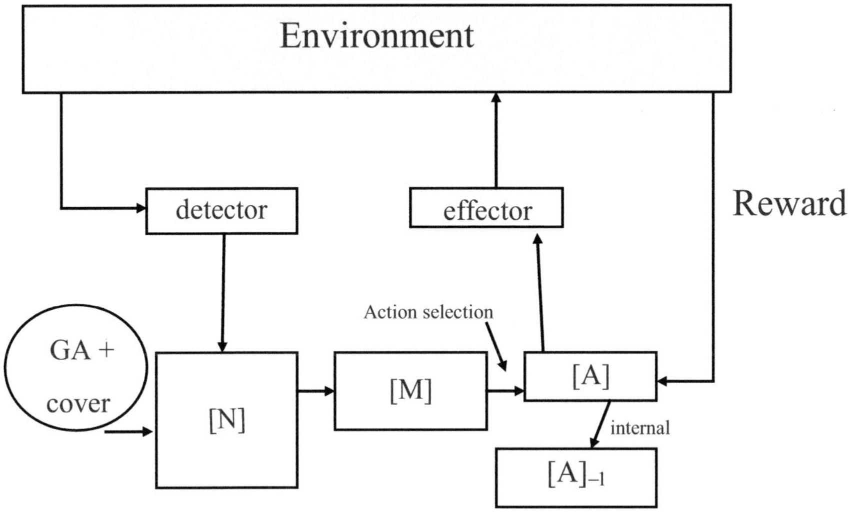

Fig. 2. Schematic of ZCS

by their number. Finally, a tax is imposed on the members of [M] that do not belong to [A] on each time-step in order to encourage exploitation of the fitter classifiers. That is, all matching rules not in [A] have their fitnesses reduced by factor τ thereby reducing their chance of being selected on future cycles. Wilson considered this technique provisional and suggested there were better approaches to controlling exploration. The effective update of action sets is thus:

$$\text{fitness}\ ([A]) \leftarrow \text{fitness}\ ([A]) + \beta\ [\text{Reward} + \gamma\ \text{fitness}\ ([A]_{+1}) - \text{fitness}\ ([A])]$$

where $0 \leq \beta \leq 1$ is a learning rate constant. Wilson noted that this is a change to Holland's formalism since specificity is not considered explicitly through bidding and pay-back is discounted by $1 - \gamma$ on each step. ZCS employs two discovery mechanisms, a steady state GA and a covering operator. On each time-step there is a probability p of GA invocation. When called, the GA uses roulette wheel selection to determine two parent rules based on fitness. Two offspring are produced via mutation and crossover. The parents donate half their fitness to their offspring who replace existing members of the population. The deleted rules are chosen using roulette wheel selection based on the reciprocal of fitness. The cover heuristic is used to produce a new rule with an appropriate condition to the current state and a random action when a match-set appears to contain low quality rules, or when no rules match an input.

When ZCS was first presented, results from its use indicated it was capable of good, but not optimal, performance [13, 50]. More recently, it has been shown that ZCS is capable of optimal performance, at least in a number of well-known test problems, but appears to be particularly sensitive to some of its parameters [7].

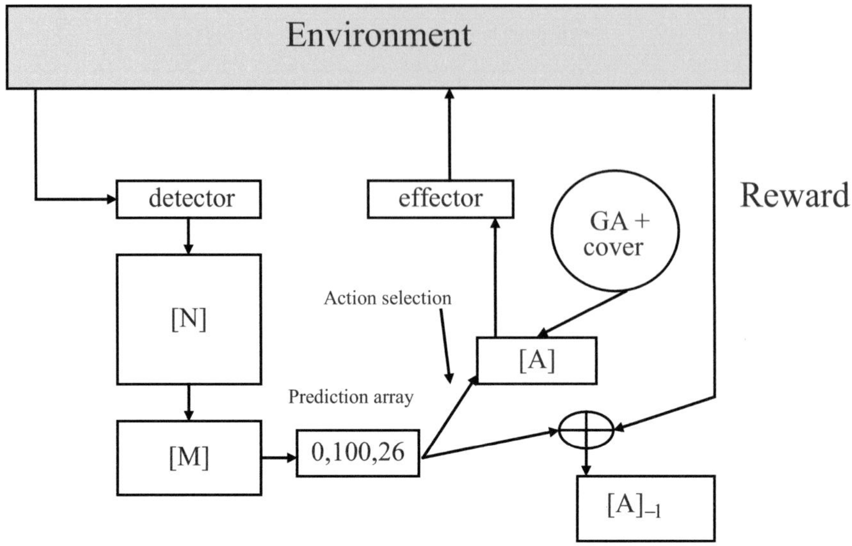

Fig. 3. Schematic of XCS

4 Wilson's XCS

The most significant difference between XCS (Fig. 3) and most other LCS (e.g., ZCS) is that rule fitness for the GA is not based on payoff received (P) by rules but on the accuracy of predictions (p) of payoff. Hence, XCS has been termed an accuracy-based LCS, in contrast to earlier systems which were for the most part strength-based (also called payoff-based systems). The intention in XCS is to form a complete and accurate mapping of the problem space (rather than simply focusing on the higher payoff niches in the environment) through efficient generalizations. In RL terms, XCS learns a value function over the complete state/action space. In this way, XCS makes the connection between LCS and reinforcement learning clear and represents a way of using traditional RL on complex problems where the number of possible state-action combinations is very large (other approaches have been suggested, such a neural networks – see [46] for an overview).

XCS shares many features with ZCS, and inherited its niche GA, deletion scheme and an interest in accuracy from Booker's GOFER-1 [4].

On each time step a match set is created. A system prediction is then formed for each action in [M] according to a fitness-weighted average of the predictions of rules in each [A]. The system action is then selected either deterministically or randomly (usually 0.5 probability per trial). If [M] is empty covering is used.

Fitness reinforcement in XCS consists of updating three parameters, ε, p and F for each rule in the current [A]; the fitness is updated according to the relative accuracy of the rule within the set in five steps:

(a) Each rule's error is updated: $\varepsilon_j = \varepsilon_j + \beta(|P - p_j| - \varepsilon_j)$ where as in ZCS $0 \leq \beta \leq 1$ is a learning rate constant.
(b) Rule predictions are then updated: $p_j = p_j + \beta(P - p_j)$
(c) Each rule's accuracy κ_j is determined: $\kappa_j = \alpha(\varepsilon_0/\varepsilon)^{\nu}$ or $\kappa_j = 1$ where $\varepsilon < \varepsilon_0$ where ν, α and ε_0 are constants controlling the shape of the accuracy function.
(d) A relative accuracy κ_j' is determined for each rule by dividing its accuracy by the total of the accuracies in the action set.
(e) The relative accuracy is then used to adjust the classifier's fitness F_j using the moyenne adaptive modifee (MAM) procedure: If the fitness has been adjusted $1/\beta$ times, $F_j = F_j + \beta(\kappa_j' - F_j)$. Otherwise F_j is set to the average of the values of κ_j' seen so far.

In short, in XCS fitness is an inverse function of the error in reward prediction, with errors below ε_0 not reducing fitness. The maximum $P(a_i)$ of the system's prediction array is discounted by a factor γ and used to update rules from the previous time step. Thus XCS exploits a form of Q-learning [48] in its reinforcement procedure, whereas Holland's 1986 system and ZCS both use a form of TD(0) (as noted in [46]).

The GA acts in action sets [A], i.e., niches. Two rules are selected based on fitness from within the chosen [A]. Rule replacement is global and based on the estimated size of each action set a rule participates in with the aim of balancing resources across niches. The GA is triggered within a given action set based on the average time since the members of the niche last participated in a GA (after [5]).

XCS is more complex than ZCS but results from its use in a number of areas have been impressive. Wilson originally demonstrated results on the Boolean multiplexer function and a maze problem [51]. Early on Kovacs emphasised its ability to learn complete, accurate, and minimal representations of Boolean functions [33]. XCS has since been widely adopted in the LCS community; the majority of contributions to a recent volume on applications of LCS [6] used XCS. An algorithmic description of XCS can be found in [11], while further details of XCS can be found in [10].

5 Pittsburgh-Style LCS

The previously described forms of LCS operate under the principle of the GA population as whole forming a cooperative set of rules for the given task. Another approach is to create a more standard GA population in which each individual represents a *complete* set of rules. This approach is known as the Pittsburgh-style LCS after its development by Stephen Smith [45] at the University of Pittsburgh, USA.

Typically, rules are again of the form "IF state THEN action" but there are no associated rule specific parameters. Rather utility is assigned to the complete set of rules once they have attempted the given task, this representing the fitness metric under a standard GA. Recombination and mutation act over the set of concatenated rules, i.e., the rules in the set form one linear string of symbols.

Action selection is typically numerosity-based wherein, for a given input, an [M] is formed and the number of rules within each [A] used to determine the output, e.g., under roulette wheel selection.

A brief overview of selected works on LCS in data mining now follows. We concentrate on pre-ZCS and XCS systems in order to complement the remaining chapters of this text.

6 Previous Research on LCS in Data Mining

Data Mining is an overarching term under which various procedures to elicit information from data may be found, including:

- Data Extraction – the collation of data from one or more sources.
- Data Cleansing – the identification and treatment of erroneous or missing datum.
- Data Reduction – the removal of features which are insufficiently correlated to the given task.
- Data Modelling – the discovery of patterns in the data.
- Model Interpretation – identification of the discovered patterns.
- Model Application – use of the identified patterns, e.g., for future predictions.

Machine Learning techniques have shown themselves to be extremely useful in data reduction and modelling, with some utility in the other data mining procedures such as model interpretation (although this can be a somewhat subjective process as different users may find different patterns more interesting than others, e.g., outliers vs. majority).

Goldberg [21] was the first to apply Holland's LCS to a real-world problem – gas pipeline control. Here the system received hourly readings from various sensors around a network, such as flow rates and pressures, and was required to deduce whether a leak was occurring and to set appropriate control of the pipeline inflow. Using a relatively small rule-base of 60 rules and a message list of size 5, Goldberg was able to train the LCS to become an effective controller after around 1,000 days of simulated use. Other early applications of Holland's system include space vessel power management [23] and modelling economic markets [36].

Following Wilson's [49] early demonstration that a version of Holland's LCS – termed BOOLE – could learn Boolean functions effectively their use for classification has become the most common application of LCS in data mining. Bonelli and Parodi [3] altered the reinforcement update of BOOLE in their system "Newboole" to penalize incorrect responses as well as reward correct ones. They showed roughly equal performance to the rule inducer CN2 [12] and a traditional neural network on three well-known data sets. They also considered the issue of rule compaction for knowledge discovery in LCS.

Riolo [40], without the use of a GA, showed human-like behaviour on a two-class discrimination task using Holland's LCS. More recently, Hartley [25] has reported similar behaviour for XCS, showing a closer agreement to humans in certain classes of problem than Newboole.

Sen [44] showed how incorporating rule accuracy, in the form of percentage correct over total number of matches, into the action selection scheme of Newboole improves performance. Using versions of the well-known MONKS problem and another data set he showed better classification accuracy than a number of popular rule induction techniques of the time, such as ID3 [39], CN2, and mFOIL [35]. Saxon and Barry [43] have used XCS on the MONKS problem.

Frey and Slate [19] used the percentage correct metric noted above as the fitness measure in a version of Holland's LCS. That is, they used a specific form of accuracy for reproduction, in a similar vein to XCS, and reported good performance on a standard letter recognition task with the modification.

Greene and Smith (e.g., [24]) developed a version of Holland's algorithm – COGIN – wherein the search is constrained to cover all training set data. The GA is modified to have random selection and competition for replacement based on the number of training examples correctly matched and the current occupation of the niches, i.e., training examples, other rules currently cover. This approach is shown to be competitive with C4.5.

DeJong, Spears and Gordon (e.g., [15]) presented early demonstrations of the ability of the Pittsburgh-style systems for data mining with their system GABIL. Using a breast cancer data set they reported favourable performance against C4.5, another ID3 derivative, and two other techniques which form modified DNF hypotheses. In particular, they showed how biased mutation operators which alter the logical relationship between chosen features can greatly improve performance.

Giordana and Neri (e.g., [20]) developed a hybrid Pittsburgh-Holland approach to evolve Horn clauses through use of a spatially distributed GA. Here each node represents a conjunction and the niching effect of the distribution enables the maintenance of multiple rule types which together solve a problem. Their system, named REGAL, was shown to be comparable to ID3 and C4.5 on a number of well-known test datasets.

Robertson and Riolo [41] presented a version of Holland's system for a number of letter sequence prediction tasks. Using the traditional windowing approach they report optimality for some tasks and near optimality on alphabet prediction. Extensions to consider using the internal memory list mechanism only was also shown possible, although a number of extra heuristics were added.

Federman and Dorchak [17] used a version of Goldberg's Simple Classifier System [22] which is much like Wilson's BOOLE, to predict the next note in a simple children's songs. They describe a correlation between prediction accuracy of the LCS and an information theory metric; perhaps unsurprisingly, the more information contained in a melody, the easier it is for the LCS to predict the sequence. This was shown to be true for the three rule representation schemes tried.

As discussed thus far, the rule representation of LCS means that the action is not a direct function of the input. Valenzuela-Rendon [47] introduced a fuzzy set rule representation scheme for LCS which has been used both for classification tasks (e.g., [38]) and, in the Pittsburgh-style, for numerical prediction (e.g., [13]). In the latter work, Cordon et al. [14] report superior performance to a number of non-linear regression methods and an artificial neural network approach. Ahluwalia and Bull [1] presented a scheme wherein each action is represented by an evolving arithmetic LISP S-expression, i.e., of the form used in Genetic Programming (GP) [34]. Although they applied it to feature extraction in conjunction with the k-nearest-neighbours algorithm, it could equally be applied to regression problems. More recently, Wilson [53] introduced a prediction estimation mechanism into XCS – termed XCSF. Weight vectors are added to each rule to enable piecewise-linear approximation based on the input. Bull and O'Hara [9] show how a neural network representation scheme can be used within XCS and demonstrate its use on both discrete action and prediction tasks. More recently, they have included the use of established gradient descent techniques for learning connection weights in conjunction with the GA search to improve accuracy [37]. The general scheme for using local search heuristics in conjunction with the GA in XCS was introduced by Wyatt and Bull [55] in their work on using XCS to classify continuous-valued problem spaces.

7 Learning Classifier Systems in Data Mining: An Overview

The rest of this book describes recent research on the use of LCS in the main areas of machine learning data mining: classification, clustering, time-series and numerical prediction, feature selection, ensembles, and knowledge discovery.

Jaume Bacardit et al. – Data Mining in Proteomics with Learning Classifier Systems. Protein structure prediction is a well-known and notoriously difficult problem whose solution offers very real benefits. This contribution describes the use of a Pittsburgh-style system to the problem and highlights the benefits from the human-readability of LCS solutions.

William Browne – Improving Evolutionary Computation Based Data Mining for the Process Industry: The Importance of Abstraction. This contribution begins by describing the application of a Holland-style system to data gathered from a steel hot strip mill. The author then suggests that mechanisms for higher levels of abstraction are needed for the real benefits of such machine learning to be seen in many domains.

Hai Dam et al. – Distributed Learning Classifier Systems. For some time now it has been recognized that using multiple classifiers, so-called ensemble machines (e.g., [26]), can prove highly effective. No one technique will outperform all others on all problems, hence the principle of ensemble machines is to combine the output of several methods to find an overall solution that utilises the strength of the constituents and compensates for their individual weaknesses. This contribution presents the use of LCS in ensembles in a truly distributed framework together with mechanisms to exploit their population-based search characteristics.

Faten Kharbat et al. – Knowledge Discovery from Medical Data: An Empirical Study with XCS. As noted above, XCS has proven to be a particularly effective data miner. This contribution describes the use of XCS to mine breast cancer data obtained from a UK health trust and its improved performance, both in terms of accuracy and the interestingness of its learned rules, compared to C4.5.

Albert Orriols-Puig and Ester Bernadó-Mansilla – Mining Imbalanced Data with Learning Classifier Systems. The class imbalance problem can be defined as a problem encountered by any inductive learning system in domains for which one class is under-represented and which assume a balanced class distribution in the training data. For a two-class problem, the class defined by the smaller set of examples is referred to as the minority class while the other class is referred to as the majority class. This contribution considers how XCS can be modified to consider this problem automatically, i.e., such that no extra forms of data manipulation, as typically used, can be avoided.

Avinash Gandhe et al. – XCS for Fusing Multi-Spectral Data in Automatic Target Recognition. This contribution describes the use of XCS for both classification and as a pre-processor for classification. That is, XCS as a feature selection approach is demonstrated on an image processing task.

Christopher Stone and Larry Bull – Foreign Exchange Trading Using a Learning Classifier System. The fact that LCS learn incrementally means they are particularly suited to on-line applications such as time-series prediction. This contribution describes a version of ZCS applied to an on-line trading task and is shown to be competitive with an approach which learns off-line in the traditional batch training mode.

Kreangsak Tamee et al. – Towards Clustering with Learning Classifier Systems. The LCS paradigm is also applicable to unsupervised learning tasks. This contribution describes modifications made to XCS to enable the identification of clusters within data sets without the prior definition of how many clusters are expected.

Albert Orriols-Puig et al. – Comparison of Several Genetic-Based Supervised Learning Systems. As noted above, it is possible to use a number of representation schemes within LCS. This contribution compares both Pittsburgh and Michigan systems using fuzzy logic representations to a number of well-known techniques.

8 Summary

Just over 30 years after Holland first presented the outline for Learning Classifier System paradigm, the ability of LCS to solve complex real-world problems is becoming clear. In particular, their capability for rule induction in data mining has sparked renewed interest in LCS. This article has given a brief introduction to LCS and previous studies of their use for data mining. The rest of the book brings together work by a number of individuals who are demonstrating their good performance in a variety of domains.

Acknowledgements

Thanks to everyone involved in this edited collection: Professor Kacprzyk for agreeing to publish the book in his series and the authors without whose efforts there would be no book.

References

1. Ahluwalia, M. & Bull, L. (1999) Coevolving functions in genetic programming: classification using K-nearest-neighbour. In: W. Banzhaf, J. Daida, A.E. Eiben, M.H. Garzon, V. Honavar, M. Jakiela & R.E. Smith (eds) *GECCO-99: Proceedings of the Genetic and Evolutionary Computation Conference*, pp. 947–952. Morgan Kaufmann, Los Altos, CA
2. Bernadó, E. Llora, X. & Garrell, J.M. (2002) XCS and GALE: a comparative study of two learning classifier systems on data mining. In: P.L. Lanzi, W. Stolzmann & S.W. Wilson (eds) *Advances in Learning Classifier Systems*, pp. 115–132, LNAI 2321. Springer, Berlin Heidelberg New York
3. Bonelli, P. & Parodi, A. (1991) An efficient classifier system and its experimental comparison with two representative learning methods on three medical domains. In: R.K. Belew & L.B. Booker (eds) *Proceedings of the 4th International Conference on Genetic Algorithms*, pp. 288–295. Morgan Kaufmann, Los Altos, CA
4. Booker, L. (1982) Intelligent Behavior as an Adaptation to the Task Environment. Ph.D. Thesis, the University of Michigan, USA
5. Booker, L. (1989) Triggered rule discovery in classifier systems. In: J.D. Schaffer (ed.) *Proceedings of the Third International Conference on Genetic Algorithms*, pp. 265–274. Morgan Kaufmann, Los Altos, CA
6. Bull, L. (2004) (ed.) *Applications of Learning Classifier Systems.* Springer, Berlin Heidelberg New York
7. Bull, L. & Hurst, J. (2002) ZCS redux. *Evolutionary Computation* 10(2): 185–205

8. Bull, L. & Kovacs, T. (2005) (eds) *Foundations of Learning Classifier Systems.* Springer, Berlin Heidelberg New York
9. Bull, L. & O'Hara, T. (2002) Accuracy-based neuro and neuro-fuzzy classifier systems. In: W.B. Langdon, E. Cantu-Paz, K. Mathias, R. Roy, D. Davis, R. Poli, K. Balakrishnan, V. Honavar, G. Rudolph, J. Wegener, L. Bull, M.A. Potter, A.C. Schultz, J.F. Miller, E. Burke & N. Jonoska (eds) *GECCO-2002: Proceedings of the Genetic and Evolutionary Computation Conference*, pp. 905–911. Morgan Kaufmann, Los Altos, CA
10. Butz, M. (2005) *Rule-Based Evolutionary Online Learning Systems.* Springer, Berlin Heidelberg New York
11. Butz, M. & Wilson, S.W. (2001) An algorithmic description of XCS. In: P.L. Lanzi, W. Stolzmann & S.W. Wilson (eds) *Advances in Learning Classifier Systems: Proceedings of the Third International Conference – IWLCS2000*, pp. 253–272. Springer, Berlin Heidelberg New York
12. Clark, P. & Niblett, P. (1987) Induction in noisy domains. In: I. Bratko & N. Lavrac (eds) *Progress in Machine Learning*, pp. 11–30. Sigma, Bled, Yugoslavia
13. Cliff, D. & Ross, S. (1995) Adding temporary memory to ZCS. *Adaptive Behavior* 3(2): 101–150
14. Cordon, O. Herrera, F. & Sanchez, L. (1999) Solving electrical distribution problems using hybrid evolutionary data analysis techniques. *Applied Intelligence* 10(1): 5–24
15. DeJong, K. Spears, W. & Gordon, D. (1993) Using genetic algorithms for concept learning. *Machine Learning* 13: 161–188
16. Eiben, A. & Smith, J. (2003) *Introduction to Evolutionary Computing.* Springer, Berlin Heidelberg New York
17. Federman, F. & Dorchak, S.F. (1997) Information theory and NEXTPITCH, a learning classifier system. In: T. Baeck (ed.) *Proceedings of the Seventh International Conference on Genetic Algorithms*, pp. 442–448. Morgan Kaufmann, Los Altos, CA
18. Forrest, S. & Miller, J.H. (1991) Emergent behaviour in classifier systems. In: S. Forrest (ed.) *Emergent Computation*, pp. 213–227. MIT Press, Cambridge, MA
19. Frey, P. & Slate, D. (1991) Letter recognition using Holland-style adaptive classifiers. *Machine Learning* 6: 161–182
20. Giordana, A. & Neri, F. (1995) Search-intensive concept induction. *Evolutionary Computation* 3: 375–416
21. Goldberg, D.E. (1983) Computer-aided gas pipeline operation using genetic algorithms and rule-learning. Ph.D. Thesis, University of Michigan
22. Goldberg, D.E. (1989) *Genetic Algorithms in Search, Optimization, and Machine Learning.* Addison-Wesley, Reading, MA
23. Goodloe, M. & Graves, S.J. (1988) Improving performance of an electric power expert system with genetic algorithms. In: *Proceedings of the 1st International Conference on the Applications of Artificial Intelligence and Expert Systems.* IEA/AIE-88, pp. 298–305
24. Greene, D.P. & Smith, S.F. (1994) Using coverage as a model building constraint in learning classifier systems. *Evolutionary Computation* 2(1): 67–91
25. Hartley, A. (1999) Accuracy-based fitness allows similar performance to humans in static and dynamic classification. In: W. Banzhaf, J. Daida, A.E. Eiben,

M.H. Garzon, V. Honavar, M. Jakiela & R.E. Smith (eds) *GECCO-99: Proceedings of the Genetic and Evolutionary Computation Conference*, pp. 266–273. Morgan Kaufmann, Los Altos, CA
26. Ho, T. Hull, J.J. & Srihari, S.N. (1994) Decision combination in multiple classifier systems. *IEEE Trans on PAMI* 16(1): 66–75
27. Holland, J.H. (1975) *Adaptation in Natural and Artificial Systems.* University of Michigan Press, Ann Arbor
28. Holland, J.H. (1976) Adaptation. In: R. Rosen & F.M. Snell (eds) *Progress in Theoretical Biology*, 4. Plenum, New York
29. Holland, J.H. (1980) Adaptive algorithms for discovering and using general patterns in growing knowledge bases. *International Journal of Policy Analysis and Information Systems* 4(3): 245–268
30. Holland, J.H. (1986). Escaping brittleness: the possibilities of general-purpose learning algorithms applied to parallel rule-based systems. In: R.S. Michalski, J.G. Carbonell & T.M. Mitchell (eds) *Machine learning, an artificial intelligence approach.* Morgan Kaufmann, Los Altos, CA
31. Holland, J.H. & Reitman, J.H. (1978) Cognitive systems based in adaptive algorithms. In: D.A. Waterman & F. Hayes-Roth (eds) *Pattern-Directed Inference Systems.* Academic, New York
32. Holland, J.H. Holyoak, K.J. Nisbett, R.E. & Thagard, P.R. (1986) *Induction: Processes of Inference, Learning and Discovery.* MIT Press, Cambridge, MA
33. Kovacs, T. (1997) XCS classifier system reliably evolves accurate, complete and minimal representations for Boolean functions. In: R. Roy, P. Chawdhry & R. Pant (eds) *Soft Computing in Engineering Design and Manufacturing*, pp. 59–68. Springer, Berlin Heidelberg New York
34. Koza, J.R. (1994) *Genetic Programming.* MIT Press, Cambridge, MA
35. Lavarac, N. & Dzeroski, S. (1994) *Inductive Logic Programming: Techniques and Applications.* Ellis Horwood, Chichester, UK
36. Marimon, R. McGrattan, E. & Sargent, J. (1990) Money as a medium of exchange in an economy with artificially intelligent agents. *Journal of Economic Dynamics and Control* 14: 329–373
37. O'Hara, T. & Bull, L. (2005) A memetic accuracy-based neural learning classifier system. In: *Proceedings of the IEEE Congress on Evolutionary Computation*, pp. 2040–2045. IEEE, New York
38. Pena-Reyes, C. & Sipper, M. (1999) A fuzzy-genetic approach to breast cancer diagnosis. *Artificial Intelligence in Medicine* 17(2): 155
39. Quinlan, J.R. (1986) Induction of decision trees. *Machine Learning* 1: 18–106
40. Riolo, R. (1991) Modeling simple human category learning with a classifier system. In: L. Booker & R.K. Belew (eds) *Proceedings of the Fourth International Conference on Genetic Algorithms*, pp. 324–333. Morgan Kaufmann, Los Altos, CA
41. Robertson, G. & Riolo, R. (1988) A tale of two classifier systems. *Machine Learning* 3: 139–159
42. Samuel, A.L. (1959) Some studies in machine learning using the game of checkers. *IBM Journal of Research and Development* 3: 211–229
43. Saxon, S. & Barry, A. (2000) XCS and the Monk's problems. In: P.-L. Lanzi, W. Stolzmann & S.W. Wilson (eds) *Learning Classifier Systems: From Foundations to Applications*, pp. 223–242. Springer, Berlin Heidelberg New York

44. Sen, S. (1993) Improving classification accuracy through performance history. In: S. Forrest (ed.) *Proceedings of the Seventh International Conference on Genetic Algorithms*, p. 652. Morgan Kaufmann, Los Altos, CA
45. Smith, S.F. (1980) *A Learning System Based on Genetic Adaptive Algorithms.* Ph.D. Thesis, Univ. Pittsburgh, USA
46. Sutton, R. & Barto, A. (1998) *Reinforcement Learning.* MIT Press, Cambridge, MA
47. Valenzuela-Rendon, M. (1991) The fuzzy classifier system: a classifier system for continuously varying variables. In: *Proceedings of the Fourth International Conference on Genetic Algorithms*, pp. 346–353. Morgan Kaufmann, Los Altos, CA
48. Watkins, C.J. (1989) Learning from Delayed Rewards. Ph.D. Thesis, Cambridge University
49. Wilson, S.W. (1987) Classifier systems and the animat problem. *Machine Learning* 2: 199–228
50. Wilson, S.W. (1994) ZCS: a zeroth-level classifier system. *Evolutionary Computation* 2(1): 1–18
51. Wilson, S.W. (1995) Classifier fitness based on accuracy. *Evolutionary Computation* 3(2): 149–176
52. Wilson, S.W. (1998) Generalization in the XCS classifier system. In: Koza et al. (eds) *Genetic Programming 1998: Proceedings of the Third Annual Conference*, pp. 322–334. Morgan Kaufmann, Los Altos, CA
53. Wilson, S.W. (2001) Function approximation with a classifier system. In: W.B. Langdon, E. Cantu-Paz, K. Mathias, R. Roy, D. Davis, R. Poli, K. Balakrishnan, V. Honavar, G. Rudolph, J. Wegener, L. Bull, M.A. Potter, A.C. Schultz, J.F. Miller, E. Burke & N. Jonoska (eds) *GECCO-2001: Proceedings of the Genetic and Evolutionary Computation Conference*, pp. 974–981. Morgan Kaufmann, Los Altos, CA
54. Wilson, S.W. & Goldberg, D.E. (1989) A critical review of classifier systems. In: *Proceedings of the 3rd International Conference on Genetic Algorithms*, pp. 244–255. Morgan Kaufmann, Los Altos, CA
55. Wyatt, D. & Bull, L. (2004) A memetic learning classifier system for describing continuous-valued problem spaces. In: N. Krasnagor, W. Hart & J. Smith (eds) *Recent Advances in Memetic Algorithms*, pp. 355–396. Springer, Berlin Heidelberg New York

Data Mining in Proteomics with Learning Classifier Systems

Jaume Bacardit[1], Michael Stout[1], and Jonathan D. Hirst[2], and Natalio Krasnogor[1]

[1] Automated Scheduling, Optimization and Planning research group, School of Computer Science and IT, University of Nottingham, Jubilee Campus, Wollaton Road, Nottingham, NG8 1BB, UK, jqb@cs.nott.ac.uk, mqs@cs.nott.ac.uk, nxk@cs.nott.ac.uk

[2] School of Chemistry, University of Nottingham, University Park, Nottingham NG7 2RD, UK, jonathan.hirst@nottingham.ac.uk

Summary. The era of data mining has provided renewed effort in the research of certain areas of biology that for their difficulty and lack of knowledge were and are still considered unsolved problems. One such problem, which is one of the fundamental open problems in computational biology is the prediction of the 3D structure of proteins, or protein structure prediction (PSP). The human experts, with the crucial help of data mining tools, are learning how protein fold to form their structure, but are still far from providing perfect models for all kinds of proteins. Data mining and knowledge discovery are totally necessary in order to advance in the understanding of the folding process. In this context, Learning Classifier Systems (LCS) are very competitive tools. They have shown in the past their competence in many different data mining tasks. Moreover, they provide human-readable solutions to the experts that can help them understand the PSP problem. In this chapter we describe our recent efforts in applying LCS to PSP related domains. Specifically, we focus in a relevant PSP subproblem, called Coordination Number (CN) prediction. CN is a kind of simplified profile of the 3D structure of a protein. Two kinds of experiments are described, the first of them analyzing different ways to represent the basic composition of proteins, its primary sequence, and the second one assessing different data sources and problem definition methods for performing competent CN prediction. In all the experiments LCS show their competence in terms of both accurate predictions and explanatory power.

1 Introduction

The prediction of the 3D structures of proteins is both a fundamental and difficult problem in computational biology. The usual approach to solve this problem is to use a divide-and-conquer approach and thus predict some specific attributes of a protein native structure, such as the secondary structure, solvent accessibility or coordination number. Accurate predictions of these

J. Bacardit et al.: *Data Mining in Proteomics with Learning Classifier Systems*, Studies in Computational Intelligence (SCI) **125**, 17–46 (2008)
www.springerlink.com

subproblems and proper understanding of the contribution of each subproblem and the rationale behind these predictions is crucial to integrating them successfully into a final 3D protein structure prediction (PSP).

Learning Classifier Systems (LCS) [1,2] are a class of evolutionary computation based machine learning techniques that could be used to tackle these issues in PSP. As they have shown in the past their competence on data mining problems [3, 4] using diverse LCSs such as XCS [5], GALE [6] or GAssist [7]. Importantly, they provide human-readable and highly interpretable solutions to the prediction problem, usually rule sets. The understanding of these solutions can lead to improvements in the way the information is represented and also more efficient integration of them into the final 3D structure prediction.

This chapter shows some competitive advantages of LCS over other techniques and also some of the challenges facing LCSs as they are applied to mining PSP datasets. Our chapter collects our recent research [8–10] using GAssist [7], a recent Pittsburgh approach LCS [2]. GAssist was applied to a PSP problem called coordination number (CN), which is defined as the prediction, for a given residue, of the number of residues from the same protein that are in contact with it. Two residues are said to be in contact when the distance between the two is below a certain threshold. The CN feature is a simplified profile of the density of the 3D structure of a protein, and therefore can be helpful in constraining the vast search space of the full PSP problem.

The chapter is divided in two parts. In the first part we analyze some alternative representations for the most basic form of information of a protein: its primary sequence. The primary sequence of a protein is a chain formed by 20 possible types of amino acids. Therefore the simplest way of representing a protein is by a string of a 20-letter alphabet. Moreover, these amino acids can be clustered based on physical and chemical properties, which lead to more simplified alphabets of representing the primary chain. The HP alphabet (hydrophobic/polar) is perhaps the best known example of reduced alphabets. This simplification is usually combined with a reduction of the number of spatial degrees of freedom by restricting the atom or residue locations to those of a lattice [11,12]. We compare some of these representations for coordination number prediction and test them using GAssist as well as other machine learning techniques.

In the second part of the chapter we show our evaluation of various types of input information, class definitions and learning algorithms applied to coordination number prediction of real proteins. Our aim is to perform a rigorous evaluation of the contribution of various kinds of information and problem definitions towards predicting CN and also analyze the explanatory power that LCSs can offer in this dataset. All the reported experiments show how LCSs can perform competitively with other learning techniques. We also show that the solutions obtained are human-readable and have rich explanatory power. This is another important advantage of GAssist as the biologists are not only interested in the quality of the predictions, but also the reasons behind them.

The rest of the chapter is structured as follows: First, Sect. 2 will contain background information and related work about proteins, CN prediction and HP lattice models. Next, Sect. 3 will describe the main characteristics of GAssist, our machine learning system. Section 4 will detail the experimental procedure. The results of applying this experimental procedure will be reported in Sect. 5. Next, Sect. 6 will discuss the results presented in the previous section and, finally, Sect. 7 will describe the conclusions and further work.

2 Problem Definition

2.1 Protein Structure and Coordination Number Prediction

Proteins are heteropolymer molecules constructed as a chain of amino acids of 20 different types. This string of amino acids is known as the primary sequence. In the native state, the chain folds to create a 3D structure. The primary sequence arranges itself into secondary structure, consisting of local structures such as alpha helices, beta sheets or coils. These local structures can group in several conformations or domains forming a tertiary structure. Secondary and tertiary structure may form concomitantly. The final 3D structure of a protein consists of one or more domains. In this context, the coordination number of a certain residue is a profile of the end product of the folding process indicating the number of other residues that end up near the target residue. Some of these contacts can be close in the protein chain but some other can be quite far apart, trivial contacts such as those with the immediate neighbour residues are ignored. Figure 1 contains a graphical representation of the CN of a residue in an alpha helix, given a minimum chain separation (ignored trivial contacts) of two. In this example, the CN of the target residue is two.

This problem is closely related to contact map (CM) prediction that seeks to predict, for all possible pairs of residues of a protein, whether they are in contact or not. When the contact map is represented as a binary matrix, the CN of a residue is the count of the number of ones in the row of the map associated with that residue. Figure 2 shows the relation between the native structure of a protein, a contact map and the coordination number of a residue. It also shows how different secondary structure elements of a protein are reflected as different kind of patterns in a contact map.

There is a large literature in CN and CM prediction, in which a variety of machine learning paradigms have been used, such as linear regression [13], neural networks [14], hidden Markov models [15], a combination of self-organizing maps and genetic programming [16] and support vector machines [17].

There are two usual definitions of the distance used to determine whether or not there is contact between two residues. Some methods use the Euclidean distance between the C_α atoms of the two residues [15], while other methods

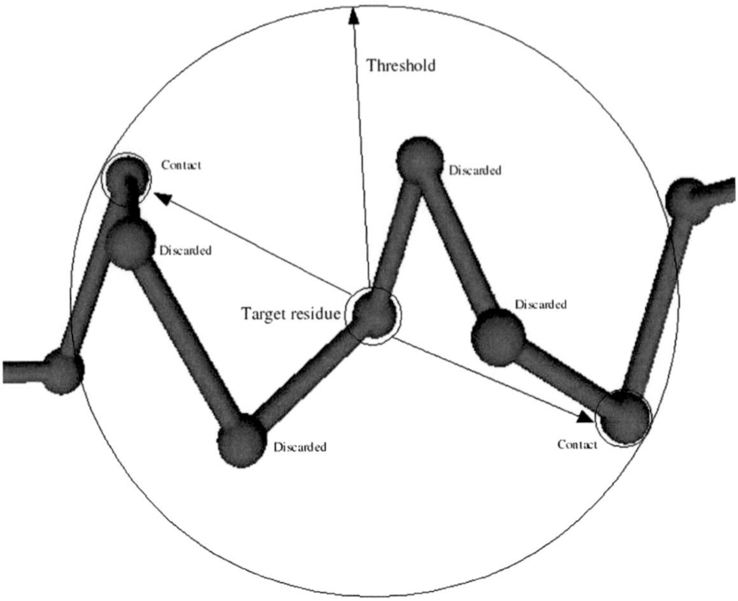

Fig. 1. Graphical representation of the CN of a residue

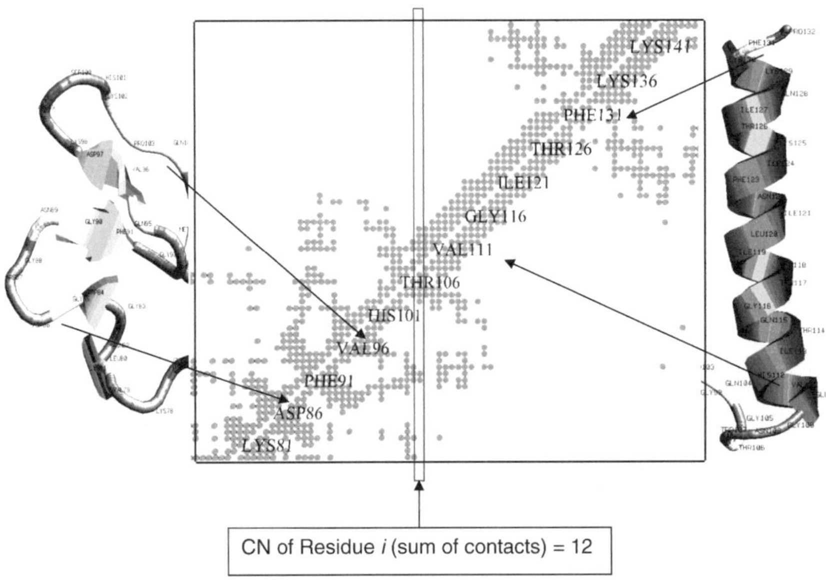

Fig. 2. Relation between a protein native structure and its contact map and the coordination number of its residues

use the C_β atom (C_α for glycine) [13]. Also, several methods discard the contacts between neighbouring residues in the primary chain by counting only contacts with a chain separation greater than a certain minimum. There are also many different distance thresholds.

Several kinds of input information are used in CN prediction, besides the amino acid (AA) type of the residues in the primary chain, such as global information of the protein chain [13], position-specific scoring matrices (PSSM) computed from multiple sequences alignments [13–17] (mainly using PSI-BLAST [18]), predicted secondary structure [14, 17], predicted solvent accessibility [14], physical characteristics of the residues [8] or sequence conservation [17]. Contact maps for any protein dataset could be easily generated through our protein structure comparison web server at http://www.procksi.net/ and used as raw data for data mining tasks.

2.2 HP Models

As protein structure prediction remains an unsolved problem, researchers have resorted to simplified protein models to try to gain understanding of both the process of folding and the algorithms needed to predict it [11, 12, 19–21]. Approaches have included fuzzy sets, cellular automata, L-systems and memetic algorithms [22–27]. One common simplification is to focus only on a representative atom of each residues (C-alpha or C-beta atoms) rather than all the atoms in the protein. A further simplification is to reduce the number of residue types to less than 20 by using representations based, for instance, on physicochemical properties such as hydrophobicity, as in the so called hydrophobic/polar (HP) models. A further simplification is to reduce the number of spatial degrees of freedom by restricting the atom or residue locations to those of a lattice [11, 12]. Lattices of various geometries have been explored, e.g., two-dimensional triangular and square geometries or three-dimensional diamond and face centered cubic [25].

In the HP model (and its variants) the 20 residue types are reduced to two classes: non-polar or hydrophobic (H) and polar (P) or hydrophilic. An n residue protein is represented by a sequence $s \in \{H, P\}^+$ with $|s| = n$. The sequence s is mapped to a lattice, where each residue in s occupies a different lattice cell and the mapping is required to be self-avoiding. The energy potential in the HP model reflects the propensity of hydrophobic residues to form a hydrophobic core.

In the HP model, optimal (i.e. native) structures are those that minimize the following energy potential:

$$E(s) = \sum_{i<j;1\leq i,j\leq n} (\Delta_{i,j}\epsilon_{i,j}) \tag{1}$$

where

$$\Delta_{i,j} = \begin{cases} 1 \text{ if } i,j \text{ are in contact and } |i-j|>1 \\ 0 \text{ otherwise} \end{cases} \tag{2}$$

Fig. 3. Representation of a protein sequence using HP alphabet and its optimal structure in the 3D cubic lattice

In the standard HP model, contacts that are HP and PP are assigned an energy of 0 and an HH contact is assigned an energy of -1. Figure 3 shows a protein sequence represented with the HP alphabet and its optimal structure when using a 3D cubic lattice.

3 The GAssist Learning Classifier System

GAssist [7] is a Pittsburgh Genetic–Based Machine Learning system descendant of GABIL [28]. The system applies an almost standard generational GA, which evolves individuals that represent complete problem solutions. An individual consists of an ordered, variable–length rule set.

We have used the GABIL [28] rule-based knowledge representation for nominal attributes and the adaptive discretization intervals (ADI) rule representation [7] for real-valued ones. Section 5 shows an example of a rule set generated by GAssist using the GABIL representation. To initialize each rule, the system chooses a training example and creates a rule that guarantees to cover this example [29].

A fitness function based on the Minimum Description Length (MDL) principle [30] is used. The MDL principle is a metric applied to a theory (a rule set here) which balances its complexity and accuracy. Our specific MDL formulation promotes rule sets with as few rules as possible as well as rules containing predicates as simple as possible. The details and rationale of this fitness formula are explained in [7].

The system also uses a windowing scheme called ILAS (incremental learning with alternating strata) [31] to reduce the run-time of the system, specially

```
Procedure Incremental Learning with Alternating Strata
Input : Examples, NumStrata, NumIterations
Initialize GA
Examples = ReorderExamples(Examples,NumStrata)
Iteration = 0
StratumSize = size(Examples)/NumStrata
While Iteration < NumIterations
        If Iteration = NumIterations − 1
          TrainingSeg = Examples
        Else
          CurrentStratum = Iteration mod NumStrata
          TrainingSeg = examples from
                        Examples[CurrentStratum · StratumSize] to
                        Examples[(CurrentStratum + 1) · StratumSize]
        EndIf
        Run one iteration of the GA with TrainingSeg
        Iteration = Iteration + 1
EndWhile
Output : Best set of rules from GA population
```

Fig. 4. Pseudocode of the incremental learning with alternating strata scheme

for dataset with hundreds of thousands of instances, as in this chapter. This mechanism divides the training set into several non-overlapping strata and chooses a different stratum at each GA iteration for the fitness computations of the individuals. ILAS empirically showed in previous experiments not only to reduce the computational cost of GAssist but also to apply generalization pressure (complementary to the one applied by the MDL-based fitness function) that helped generating more compact and accurate solutions. Figure 4 shows the pseudocode of the ILAS windowing scheme.

Finally, we have used a ensemble mechanism wrapped over GAssist to boost its performance. We generate several rule sets using GAssist with different initial random seeds and combine them as an ensemble, producing a consensus prediction using a simple majority vote. This approach is similar to Bagging [32] but simpler as, unlike Bagging, it does not need to scramble the training set to generate slightly different classifiers to combine them as an ensemble. In previous work [33] we empirically evaluated this ensemble mechanism over a set of 25 real-world datasets from the University of California at Irvine (UCI) repository [34]. On average, the ensemble obtained a test accuracy 2.5% higher than the standalone GAssist.

GAssist used its standard parameters [7] with the 1,000 iterations for the runs in the first stage, and 20,000 for the runs in the second stage, 150 strata for the ILAS windowing scheme, and ten rule sets per ensemble.

4 Experimental Framework

4.1 Experimental Battery I: Primary Sequence and Coordination Number

The first part of the experiments reported in this chapter will focus on the relation between the primary sequence attributes of a protein and CN prediction. To analyze this relation we will compare the CN prediction for simplified HP lattice model proteins (Lattice-HP) with the prediction of the same feature for real proteins using either all 20 amino acid types (Real-AA) or using only the HP representation (Real-HP). The characteristics of each dataset are summarized in Table 1 and detailed in the rest of this subsection.

Moreover, also several degrees of precision for the prediction of CN will be evaluated. This feature can be defined either as a integer variable of high cardinality or directly as a continuous variable. Therefore, in order to use classification methods for CN prediction, this feature has to be transformed into a more or less reduced finite set of labels. Thus, the experiments in this first battery contain datasets with varying degree on precision in two dimensions: the primary sequence (inputs of the classification problem) and the CN feature (output of the classification problem). Figure 5 shows these two dimensions of detail of both the primary sequence and the CN, naming some of the different possible choices of representation for both of them, marking with a dashed area the ones that have been used in this chapter. It is important to remark, however, that in some cases a larger degree of detail does not mean

Table 1. Details of the data sets used in these experiments

Name	Lattice-HP	Real-HP/Real-AA
Type	3D Cubic Lattice	Real Proteins
Number of Sequences	15	1,050
Minimum Sequence Length	27	80
Maximum Sequence Length	48	2,329
Total Hydrophobic	316	170,493
Total Polar	309	84,850
Total Residues	625	255,343

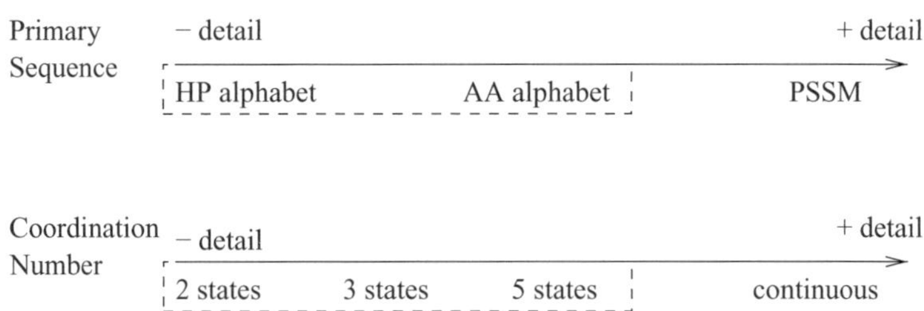

Fig. 5. Degrees of precision in each dimension of the CN prediction problem. The dashed areas mark the representation options explored in this chapter

that the problem becomes more difficult, one such case is treating the CN as a continuous variable or as a discrete one. In the first case the problem becomes a regression domain, in the second one it becomes a classification domain. Both problems are equally challenging.

Real Proteins Dataset

The datasets based on real proteins (Real-AA and Real-HP) use the CN definition proposed by Kinjo et al. [13] defined as follows. The distance used to determine contact is defined using the C_β atom (C_α for glycine) of the residues. The boundary of the sphere defined by the distance cutoff $d_c \in \Re^+$ is made smooth by using a sigmoid function. Also, a minimum chain separation of two residues is required. Formally, the CN (O_i^p) of the residue i of protein chain p is computed as:

$$O_i^p = \sum_{j:|j-i|>2} \frac{1}{1 + exp(w(r_{ij} - d_c))} \tag{3}$$

where r_{ij} is the distance between the C_β atoms of the ith and jth residues. The constant w determines the sharpness of the boundary of the sphere. A value of three for w was used for all the experiments. CN was computed using a distance cutoff of 10 Å. In previous work [9] we empirically tested different distance cutoffs, and 10 Å was the one that obtained highest accuracy.

This CN definition is real-valued. Therefore, it has to be converted into a set of finite classes so that it can be used as a classification dataset. The continuous CN domain will be discretized using the Uniform Frequency discretization algorithm [35]. Three numbers of states will tested for the experiments reported in this chapter, dividing the CN domain into two, three and five states.

The real protein dataset (Real-AA) was selected from PDB-REPRDB [36], an online server that automatically selects curated protein subsets given a set of criteria with the following conditions: less than 30% sequence identity, sequence length greater than 50, no membrane proteins, no nonstandard residues, no chain breaks, resolution better than 2 Å and having a crystallographic R factor better than 20%. Chains that had no entry in the HSSP [37] database were discarded. The final data set contains 1,050 protein chains.

Each instance of the dataset has as class the CN of a residue belonging to some of these 1,050 protein chains, and as input information we will use local information of the target residue and its closest neighbours in the chain: To generate the instance in the dataset we will use the most usual method, define a sliding window, centered initially over the first residue of a chain and that will move one residue at a time. This window is centered over the target and includes $\pm N$ residues, meaning that the CN of the target will be predicted using as inputs the primary sequence representation of the residues in the window. Windows were generated for one, two and three residues at

Table 2. Assignment of residues as hydrophobic or polar

Residue (one letter code)	Assignment
ACFGILMPSTVWY	Hydrophobic
DEHKRQN	Polar

each side of the central residue. The set was divided randomly into ten pairs of training and test set using 950 proteins for training and 100 for testing in each set. These sets act in a similar way to a ten-fold cross-validation. The proteins included in each partition are reported in http://maccl01.genes.nig.ac.jp/~akinjo/sippre/suppl/list/. This same dataset was used to generate a real protein HP sequence dataset (Real-HP) by assigning each residue a value of Hydrophobic or Polar as shown in Table 2, following Broome and Hecht [38].

HP Lattice-Based Datasets

For the Lattice-HP study, a set of structures from Hart's Tortilla Benchmark Collection (http://www.cs.sandia.gov/tech_reports/compbio/tortilla-hp-benchmarks.html) was used. This consisted of 15 structures on the simple cubic lattice, in which each residue can have up to six neighbours in the lattice, therefore the maximum CN value that a residue can have is also 6. Again, windows were generated for one, two and three residues at each side of a central residue and the CN class of the central residue assigned as the class of the instance. The instance set was partitioned into training/test sets using stratified ten-fold cross-validation. The process was repeated ten times to generate ten sets of cross-validation folds. Each reported accuracy will be, therefore, the average of one hundred values. As in the real dataset, the CN of this domain will be divided into two, three and five states using an uniform frequency discretization algorithm.

Comparison of LCS Performance

The performance of GAssist in the Lattice-HP, Real-HP and Real-AA datasets will be compared against two well known machine learning methods. We chose two C4.5 [39] and Naive Bayes [40], using the WEKA [41] implementation of both of them. Student t-tests are applied to the results of the experiments to determine, for each dataset if the best method is significantly better than the other algorithms using a confidence interval of 95%. The Bonferroni correction [42] for multiple pair-wise comparisons has been used.

4.2 Experimental Battery II: Assessment of Input Information Sources and Class Partitions for Coordination Number

The second part of the experiments reported in this chapter focuses exclusively on real proteins, assessing two different dimensions of the CN problem: (a) what is the contribution of several types of input information and (b) different criteria to define the classes of the CN domain. We will explore six different sets of input information, two class partition criteria and three different numbers of class partitions, testing in total 36 datasets.

Class Definitions and Protein Dataset

We will use again the Kinjo et al. [13] definition of CN as well as their protein dataset and training/test partitions, detailed in the previous subsection. As explained previously, the CN definition is continuous, and has to be discretized to handle the dataset as a classification domain. Previously, we used the uniform frequency (UF) unsupervised discretization to generate the class partitions. For these experiments, besides UF we will also test the uniform-length (UL) unsupervised discretization [35]. For each of them, as in the first stage experiments, we will test three numbers of states, dividing the CN domain into two, three and five states.

Explored Input Information

Six sets of input attributes are evaluated in this battery of experiments. The first set corresponds to the representation used for the experiments in the first part of the chapter: the AA type of the residues in a window around the target one. The following sets add extra information, such as global protein information or predicted characteristics of the protein. The set of input attributes are labeled *CN1* through *CN6*. This allows us to assess rigorously whether additional information is of benefit, and the degree of usefulness of each kind of extra data.

The global protein information consists of 21 real-valued attributes. The first attribute is the length of the protein chain (number of residues). The other 20 attributes contain the frequency of each AA type in the protein chain. Two types of predicted information have been used. The first is the average real-valued CN of a protein chain [13], called *PredAveCN*. This feature was predicted using GAssist itself. PredAveCN was partitioned into ten classes (Ten different states in the PredAveCN domain), using the two partition criteria (uniform length and uniform frequency) described above. This protein-wise feature was predicted from the 21 global protein attributes stated above, that is, the protein length and the frequency of appearance of the 20 AA types in the chain. The second predicted information is secondary structure of a window of residues around the target residue, using the PSI-PRED predictor [43]. This predicted information consists of two parts: a secondary structure type (helix, strand or coil) and a confidence level ([0 to 9]) of the prediction.

Table 3 summarizes the input attributes used in the datasets, and Table 4 describes which attributes are included in each sets of input information. CN3 and CN5 represent two different ways of aggregating the same source of information to CN1, either as global information or as a predicted information.

Table 3. Input attribute definitions for the tested datasets

Att. source	Description	Type	Cardinality
Len	Number of residues in a protein chain	real-valued	1 attribute
FreqRes	Frequencies of appearance of the each AA type in the protein chain	real-valued	20 attributes
AA-type	The AA type of a window of ±4 residues around the target residue	nominal	9 attributes
PredAveCN	Predicted average CN of a protein	nominal	1 attribute
PredSS	Predicted secondary structure of the ±4 residues around the target residue	nominal+real-valued	18 attributes

Table 4. Definition of the input attributes for all the used datasets

Domain	Attributes	#real-valued att.	#nominal att.	total #att.
CN1	AA-type	0	9	9
CN2	AA-type,PredSS	9	18	27
CN3	AA-type,Len,FreqRes	21	9	30
CN4	AA-type,Len,FreqRes,PredSS	30	18	48
CN5	AA-type,PredAveCN	0	10	10
CN6	AA-type,PredAveCN,PredSS	9	19	28

CN2, CN4 and CN6 add the predicted secondary structure to CN1, CN3 and CN5, respectively. Unlike battery I of experiments, Here we have only used one window size: four residues at each side of the target has been used, for both AA-type and PredSS types of input information.

Performance Measure

The accuracy metric used for these experiments is not the standard machine learning accuracy metric (#correct examples/#total examples). As is usual in the protein structure prediction field [13,43], we will take into account the fact that each example (a residue) belongs to a protein chain. Therefore, we will first compute the standard accuracy measure for each protein chain, and then average these accuracies to obtain the final, protein-wise, accuracy metric. Because different chains have different lengths, the used measure can differ from the standard accuracy. The rationale for this is to mimic the real-life situation, in which a new protein is sequenced, and researchers are interested in the predicted properties based on the entire protein sequence, independent of its length.

Comparison of LCS Performance

The performance of GAssist on these 36 datasets (6 sets of input attributes, two class definition criteria, three numbers of classes) will be compared against three other machine learning systems: C4.5 [39], a rule induction system, Naive Bayes [40], a Bayesian learning algorithm and LIBSVM [44], a support vector machine using RBF kernels. We have used the WEKA implementations [41] of both C4.5 and Naive Bayes. Student *t*-tests are applied to the results of the experiments to determine, for each dataset if the best method is significantly better than the other algorithms using a confidence interval of 95%. Again, the Bonferroni correction [42] for multiple pair-wise comparisons has been used.

5 Results

5.1 Experimental Battery I

Lattice-HP Datasets

Table 5 compares the results of two, three and five state CN predictions for a range of window sizes for the GAssist LCS, Naive Bayes and C4.5 using the Lattice-HP dataset. As the number of states is increased the accuracy decreases from around 80% to around 51% for all algorithms. For each state as the window size is increased the accuracy increases by around 0.1–0.2%. With the exception of the C4.5 algorithm which shows a decrease in accuracy with increasing window size in two and three state predictions. There were no significant differences detected in these tests and thus all learning methods showed similar performance.

Real Proteins

Table 6 compares the results of two, three and five state CN predictions on real proteins for the GAssist LCS, Naive Bayes and C4.5 using the Real-AA and Real-HP datasets.

Table 5. Lattice-HP prediction accuracies

Number of States	Algorithm	Window Size		
		1	2	3
2	GAssist	79.8 ± 4.9	80.2 ± 5.0	80.0 ± 5.3
	C4.5	80.2 ± 4.9	79.9 ± 5.0	79.7 ± 5.1
	NaiveBayes	79.8 ± 4.9	80.0 ± 4.9	80.2 ± 5.0
3	GAssist	67.4 ± 4.9	67.8 ± 4.1	67.3 ± 5.0
	C4.5	67.5 ± 4.8	67.6 ± 4.2	66.6 ± 5.0
	NaiveBayes	67.2 ± 4.6	67.3 ± 4.4	67.5 ± 4.8
5	GAssist	51.4 ± 4.6	51.3 ± 4.2	52.7 ± 5.3
	C4.5	51.7 ± 4.5	51.0 ± 4.1	52.2 ± 5.1
	NaiveBayes	51.7 ± 4.6	52.3 ± 4.3	51.9 ± 5.6

Table 6. CN prediction accuracies for the real-HP and real-AA datasets. A • means that GAssist outperformed the Algorithm to the left (5% t-test significance). A ∘ label means that the Algorithm on the left outperformed GAssist (5% t-test significance)

State	Algorithm	Window Size					
		HP Based			Residue Based		
		1	2	3	1	2	3
2	GAssist	63.6 ± 0.6	63.9 ± 0.6	64.4 ± 0.5	67.5 ± 0.4	67.9 ± 0.4	68.2 ± 0.4
	C4.5	63.6 ± 0.6	63.9 ± 0.6	64.4 ± 0.5	67.3 ± 0.4	67.5 ± 0.3	67.8 ± 0.3
	NaiveBayes	63.6 ± 0.6	63.9 ± 0.6	64.3 ± 0.5	67.6 ± 0.4	68.0 ± 0.4	68.8 ± 0.3∘
3	GAssist	44.9 ± 0.5	45.1 ± 0.5	45.6 ± 0.4	48.8 ± 0.4	49.0 ± 0.4	49.3 ± 0.4
	C4.5	44.9 ± 0.5	45.1 ± 0.5	45.8 ± 0.4	48.8 ± 0.3	48.7 ± 0.3	49.1 ± 0.3
	NaiveBayes	44.7 ± 0.5	45.2 ± 0.5	45.7 ± 0.4	49.0 ± 0.4	49.6 ± 0.5∘	50.7 ± 0.3∘
5	GAssist	29.0 ± 0.3	29.6 ± 0.5	30.1 ± 0.5	32.2 ± 0.3	32.5 ± 0.3	32.7 ± 0.4
	C4.5	29.0 ± 0.3	29.7 ± 0.4	30.4 ± 0.5	31.9 ± 0.4	31.4 ± 0.4•	31.0 ± 0.5•
	NaiveBayes	29.0 ± 0.3	29.7 ± 0.4	30.1 ± 0.5	33.0 ± 0.2∘	33.9 ± 0.3∘	34.7 ± 0.4∘

When an HP sequence representation was used, an increase in the number of states is accompanied by a decrease in accuracy from around 63–64% to around 29–30% for all algorithms. For each state, as the window size is increased the accuracy increases by around 1%. Again, no significant differences were found between the methods for the Real-HP datasets.

Using full residue information, an increase in the number of states is accompanied by a decrease in accuracy from around 68% to around 34% for all algorithms. For each state, as the window size is increased, the accuracy increases by around 0.5%, with the exception of the C4.5 algorithm which shows a decrease in accuracy with increasing window size in five state predictions. The LCS outperformed C4.5 two times and was outperformed by Naive Bayes six times.

Most interestingly, moving from HP sequence representation to full residue type sequence information only results in a 3.8% accuracy increase for two states (64.4% vs. 68.2%), 3.3% for three states (45.6% vs. 49.3%) and 2.6% for the five states class definition (30.1% vs. 32.7%).

Brief Estimation of Information Loss

In order to understand the effect of using a lower-dimensionality profile of a protein chain such as the HP model, we have computed some simple statistics on the datasets. Two measures are computed:

$$\text{redundancy} = 1 - \frac{\#\text{unique instances}}{\#\text{total instances}} \tag{4}$$

$$\text{inconsistency} = \frac{\left(\frac{\#\text{unique instances}}{\#\text{unique antecedents}}\right) - 1}{\#\text{states} - 1} \tag{5}$$

Table 7. Redundancy and inconsistency rate of the tested real-proteins datasets

States	Window Size	HP representation		AA representation	
		Redundancy	Inconsistency	Redundancy	Inconsistency
2	1	99.99%	100.000%	93.69%	90.02%
	2	99.94%	92.50%	6.14%	3.85%
	3	99.75%	81.71%	0.21%	0.05%
3	1	99.98%	96.88%	90.90%	87.01%
	2	99.92%	86.25%	4.50%	2.84%
	3	99.66%	76.00%	0.17%	0.04%
5	1	99.97%	93.75%	85.84%	81.52%
	2	99.86%	86.25%	2.97%	1.84%
	3	99.46%	74.36%	0.14%	0.03%

The redundancy metric in equation 4 illustrates the effect of reducing the alphabet and the window size: creating many copies of the same instances. The inconsistency metric in equation 5 shows how this reduction creates inconsistent instances: instances with equal input attributes (antecedent) but different class. For the sake of clarity this measure has been normalized for the different number of target states. Table 7 shows these ratios. For two-states and window size of one, the Real-HP dataset shows the most extreme case: any possible antecedent appears in the data set associated to both classes. Fortunately, the proportions of the two classes for each antecedent are different, and the system can still learn. We see how the Real-HP dataset is highly redundant and how the Real-AA dataset of window size two and three presents very low redundancy and inconsistency rate. This shows both why the window size has to be large enough and also why we have to use a rich enough primary sequence representation.

5.2 Experimental Battery II

This second part of the experiments of the chapter is also divided itself in two parts. The first explains the results of the experiments defined in Sect. 4. In the second part we perform an interpretability analysis of the results obtained by GAssist on these datasets. Some discussion follows.

Results of the Learning Experiments

We tested the selected learning systems on the 36 datasets, as summarized in Table 8. Each value is the protein-wise accuracy metric defined in previous section and averaged over the ten test sets. The t-tests applied to these results are summarized in Table 9, where each cell counts how many times the method in the row significantly outperforms the method in the column with a confidence level of 95% and the bonferroni correction.

Table 8. Accuracy of the tested systems on the CN1..CN6 datasets. A • marks methods that were significantly outperformed by GAssist, while a ◦ marks methods that significantly outperformed GAssist in that dataset. Student T-tests with 95% confidence level were applied

Dataset	System	Uniform frequency class def.			Uniform length class def.		
		2 states	3 states	5 states	2 states	3 states	5 states
CN1	GAssist	69.0 ± 0.5	50.7 ± 0.5	34.2 ± 0.3	75.9 ± 0.8	63.8 ± 0.9	46.5 ± 0.9
	Naive Bayes	68.7 ± 0.5	50.7 ± 0.6	34.5 ± 0.5	76.3 ± 0.7	64.0 ± 0.8	47.0 ± 0.8
	C4.5	68.1 ± 0.4•	49.4 ± 0.4•	30.9 ± 0.6•	75.0 ± 0.7	63.3 ± 0.9	46.1 ± 0.9
	LIBSVM	68.9 ± 0.4	51.4 ± 0.6	35.5 ± 0.6◦	77.4 ± 0.8◦	65.0 ± 0.8◦	46.9 ± 0.8
CN2	GAssist	71.0 ± 0.5	53.6 ± 0.4	35.9 ± 0.4	79.0 ± 0.7	65.8 ± 0.9	47.0 ± 0.9
	Naive Bayes	66.3 ± 0.7•	49.8 ± 0.6•	33.4 ± 0.5•	72.1 ± 0.7•	61.3 ± 1.0•	39.9 ± 0.7•
	C4.5	70.6 ± 0.6	52.8 ± 0.4•	33.6 ± 0.4•	77.9 ± 0.6•	66.7 ± 0.9	46.5 ± 1.0
	LIBSVM	72.7 ± 0.6◦	57.0 ± 0.6◦	39.0 ± 0.5◦	79.9 ± 0.6◦	69.1 ± 1.0◦	48.7 ± 0.9◦
CN3	GAssist	70.9 ± 0.5	52.6 ± 0.7	35.7 ± 0.6	77.2 ± 1.1	65.1 ± 0.9	47.0 ± 0.8
	Naive Bayes	67.7 ± 0.7•	50.4 ± 0.9•	34.1 ± 0.8•	76.2 ± 0.9	62.5 ± 1.1•	43.5 ± 1.4•
	C4.5	69.9 ± 0.5•	50.1 ± 0.7•	31.1 ± 0.6•	77.0 ± 0.9	65.1 ± 0.7	44.0 ± 0.7•
	LIBSVM	72.0 ± 0.4◦	55.3 ± 0.8◦	38.0 ± 0.5◦	79.3 ± 1.0◦	68.1 ± 0.8◦	47.2 ± 0.7
CN4	GAssist	72.7 ± 0.4	55.3 ± 0.6	37.5 ± 0.4	80.1 ± 0.8	66.9 ± 0.9	47.7 ± 0.9
	Naive Bayes	69.8 ± 0.8•	52.7 ± 0.9•	36.1 ± 0.9•	76.9 ± 1.0•	64.2 ± 0.9•	43.7 ± 1.2•
	C4.5	72.2 ± 0.4	53.4 ± 0.5•	34.0 ± 0.5•	79.1 ± 0.7	67.6 ± 0.7	45.2 ± 0.7•
	LIBSVM	75.9 ± 0.4◦	59.9 ± 0.7◦	41.9 ± 0.4◦	81.7 ± 0.7◦	71.5 ± 0.8◦	50.8 ± 0.9◦
CN5	GAssist	71.2 ± 0.5	52.9 ± 0.9	35.9 ± 0.8	77.2 ± 0.9	65.3 ± 0.8	47.1 ± 0.8
	Naive Bayes	71.5 ± 0.5	54.0 ± 0.8◦	37.3 ± 0.7◦	78.4 ± 0.8	67.2 ± 0.8◦	48.7 ± 0.7◦
	C4.5	70.3 ± 0.6	51.7 ± 0.8•	33.1 ± 0.8•	77.1 ± 1.0	65.8 ± 0.7	47.0 ± 0.8
	LIBSVM	72.0 ± 0.6◦	55.0 ± 0.8◦	37.8 ± 0.7◦	79.1 ± 0.9◦	67.9 ± 0.7◦	47.7 ± 0.9
CN6	GAssist	72.9 ± 0.4	55.9 ± 0.7	37.8 ± 0.7	80.3 ± 0.7	67.3 ± 0.8	47.8 ± 0.8
	Naive Bayes	68.5 ± 0.5•	52.0 ± 0.7•	35.1 ± 0.6•	75.2 ± 0.9•	63.9 ± 0.8•	42.8 ± 0.5•
	C4.5	72.4 ± 0.5	54.5 ± 0.6•	35.5 ± 0.6•	79.5 ± 0.7	68.4 ± 0.7◦	48.1 ± 0.8
	LIBSVM	75.8 ± 0.4◦	59.8 ± 0.6◦	41.7 ± 0.6◦	81.5 ± 0.8◦	71.3 ± 0.7◦	50.8 ± 0.8◦

Table 9. Number of times a method significantly outperforms another in the battery II of experiments, according to t-tests with 95% confidence level and Bonferroni correction for multiple comparisons

	GAssist	Naive Bayes	C4.5	LIBSVM	Times outperforming
GAssist	–	23	17	0	40
Naive Bayes	4	–	11	0	15
C4.5	1	16	–	0	17
LIBSVM	31	26	34	–	91
Times outperformed	36	65	62	0	

From the tested sets of input attributes, we can say that all the different kind of attributes sources contribute to increasing the predictive accuracy of the tested systems, as all CN2-CN6 datasets obtain higher performance than CN1. We can quantify the contribution of the predicted secondary structure information as an accuracy increase of 2-n-3% on most datasets and learning systems, comparing the performance of CN1-CN2, CN3-CN4 and CN5-CN6.

The two ways of adding global protein information to the instances, by either explicitly adding global protein descriptors or by adding the predicted protein average coordination number manage to obtain similar performance levels, by looking at the differences between CN3-CN5 and CN4-CN6. The contribution of this kind of input information to the accuracy increase is 1.5–2%.

GAssist had an average run-time ranging from 9.5 to 14 h in the CN3 dataset, while it had a run-time ranging from 0.3 to 1.1 h in the CN5 dataset. The reason of this is two-fold. First of all, GAssist has to explore a larger search space. Also, the mix or real-valued and nominal attributes requires the use of a less efficient knowledge representation. Considering this issue and the fact that the solutions generated by the CN5 datasets use less attributes than the ones generated by CN3 (therefore, more readable) it is reasonable to recommend the use of the latter kind input attribute for future experiments.

The UL class definition leads to better accuracy than the UF definition for all datasets. This reflects the capacity of the UL definition to adapt itself to the physical reality of the proteins as its criterion is based on the dimensions of the CN domain. The UF definition, a priori, may look more appropriate from a machine learning point of view, as it creates well balanced class distributions. However, it might happen that the class frontiers separate examples that are practically equal. Nevertheless, it could be worth to study the amount of information contributed by both measures. It may be possible that the UF definition, although leading to lower accuracy, provides more added value to a final 3D protein structure predictor.

Looking at the specific results of each learning method, we can observe that both GAssist and C4.5 obtain their highest accuracy in the CN6 dataset, Naive Bayes in the CN5 dataset and LIBSVM in CN4. LIBSVM achieves the best accuracy in 33 of the 36 datasets, as reflected by the t-tests, where LIBSVM outperforms the other methods in 91 of 108 times, and it is never significantly outperformed. The t-tests place GAssist in the second position of the ranking for both the number of times it outperforms C4.5 and Naive Bayes and the number of times it is outperformed by the other methods. Finally, both C4.5 and Naive Bayes perform comparably, at the bottom of the ranking.

Interpretability and Explanatory Power of GAssist Results

Table 10 summarizes two simple metrics of the solutions: the average number of rules per rule set and the average number of expressed attributes in the generated rules, that is, attributes that GAssist considered to be relevant for that rule. We see that GAssist creates compact solutions, ranging from just two rules in the CN1-2 classes - nUL dataset to 7.5 rules in the CN1 - 5 classes - UF dataset. At most, an average of 11.8 attributes were expressed in the CN4 - 2 classes-UL dataset (out of 42 attributes). In comparison, C4.5 (using pruning) sometimes generated solutions with as many as 8,000 leaves,

Table 10. Complexity measures of the GAssist solutions on the CN1..CN6 datasets. #rules = average number of rules per rule set. Exp. Att. = average number of expressed attributes per rule

Dataset	Metric	Uniform frequency class def.			Uniform length class def.		
		2 states	3 states	5 states	2 states	3 states	5 states
CN1	#rules	6.5 ± 1.1	6.4 ± 0.8	7.5 ± 0.7	2.0 ± 0.0	7.1 ± 0.6	5.4 ± 0.6
	Exp. Att.	6.6 ± 3.2	6.4 ± 3.1	6.9 ± 3.0	4.2 ± 4.2	7.2 ± 3.0	6.3 ± 3.3
CN2	#rules	6.7 ± 1.0	6.5 ± 0.7	7.1 ± 0.3	5.0 ± 0.1	5.8 ± 0.7	5.8 ± 0.7
	Exp. Att.	9.9 ± 4.7	9.3 ± 4.6	9.8 ± 4.5	11.5 ± 6.0	8.0 ± 4.3	9.5 ± 4.9
CN3	#rules	5.4 ± 0.6	5.4 ± 0.5	6.2 ± 0.4	4.1 ± 1.5	6.3 ± 0.7	5.6 ± 0.6
	Exp. Att.	7.5 ± 4.0	7.2 ± 3.9	7.7 ± 3.8	8.2 ± 4.8	6.4 ± 3.7	7.6 ± 3.9
CN4	#rules	5.9 ± 1.0	6.5 ± 0.7	6.9 ± 0.4	5.0 ± 0.2	5.7 ± 0.7	5.6 ± 0.6
	Exp. Att.	9.8 ± 5.0	9.7 ± 4.8	10.0 ± 4.7	11.8 ± 6.4	7.4 ± 4.7	9.7 ± 5.1
CN5	#rules	6.3 ± 0.9	6.6 ± 1.0	6.5 ± 0.7	2.0 ± 0.3	6.6 ± 0.6	5.6 ± 0.7
	Exp. Att.	7.3 ± 3.5	7.2 ± 3.4	7.3 ± 3.4	4.5 ± 4.5	6.8 ± 3.3	7.0 ± 3.6
CN6	#rules	6.4 ± 1.0	7.1 ± 0.7	7.0 ± 0.4	5.0 ± 0.2	6.4 ± 0.7	5.8 ± 0.7
	Exp. Att.	10.0 ± 4.9	10.3 ± 4.7	9.9 ± 4.6	13.1 ± 7.5	9.0 ± 5.4	10.3 ± 5.8

and LIBSVM used around 160,000 instances from the training set as support vectors. No simple complexity measure can be extracted from Naive Bayes.

The case of the CN1 dataset using the uniform length classes definition and two classes is especially interesting. In this dataset GAssist always generated solutions with just two rules, obtaining an average accuracy of 75.9%. One such rule set is shown below, where $AA_{\pm n}$ denotes the AA type at the position $\pm n$ in respect to the target residue, the AA type is represented using the one letter code and the symbol X is used to indicate the end of chain, for the case when the window overlaps with the beginning or the end of the protein chain:

1. If $AA_{-4} \notin \{X\}$ and $AA_{-3} \notin \{D, E, Q\}$ and $AA_{-1} \notin \{D, E, Q\}$ and $AA \in \{A, C, F, I, L, M, V, W\}$ and $AA_1 \notin \{D, E, P\}$ and $AA_2 \notin \{X\}$ and $AA_3 \notin \{D, E, K, P, X\}$ and $AA_4 \notin \{E, K, P, Q, R, W, X\}$ then class is 1
2. Default class is 0

We see two types of predicates: those stating if the AA type of a certain position of the window belongs or does not belong to a certain subset of the amino acids. When the number of AA types that the predicate for a certain residue can take includes more than ten letters, that is, half of the alphabet, GAssist generates the complementary predicate to produce a more compact solution. Therefore, all the predicates defined as $\in$ are more specific than the ones defined as $\notin$. The more specific attributes are usually also the most relevant ones, and in this rule set we only have one such predicate: the one associated to the target residue. It is reasonable to expect that the more relevant attributes are those associated directly to the residue for which we are predicting its CN.

Table 11. Expression and generality rate for the rule sets generated by GAssist for the CN1 dataset and uniform-length class definition

Window position	Expression rate	Generality rate
−4	95%	94.5% ± 4.6
−3	99%	88.1% ± 4.3
−2	57%	98.2% ± 2.5
−1	100%	84.7% ± 5.7
0	100%	39.4% ± 2.3
1	100%	83.5% ± 3.2
2	80%	96.2% ± 3.1
3	100%	78.8% ± 5.8
4	100%	78.5% ± 4.7

Table 11 contains the average number of AA types included in the predicate associated to each window position, for all the rule sets produced for this dataset, which is it is quite a good metric for the generality degree of the predicates associated to each window position. This table also reports the percentage of times that each window position was expressed in the generated rule. A non expressed attribute is irrelevant for the prediction. We observe that the window positions to the right of the target residue are more relevant than the ones to the left, and that the window positions ±2 are the most irrelevant ones. Further analysis should be performed to determine if there is a physical explanation for this issue (such as a correlation with the cycle of an alpha helix) or if it is just the effect of a GA positional bias [45].

Moreover, we can extract a simple physico–chemical explanation of such predicates: the set of AA types contained in the predicate associated to the target residue (A,C,F,I,L,M,V,W) are all hydrophobic [46]. Hydrophobic residues are usually found in the inner part of a protein in native state. Therefore it is logical that they present higher CN than the other residues, as this rule predicts high CN. This also matches with all the observations we did in battery I of experiments of this chapter.

On the other hand, from the rest of predicates of the rule set, the most frequently appearing AA types in the negated predicates are D and E, which are negatively charged. These types of AA usually appear only on the surface of the proteins, so it is sensible that they are not included in the predicates of a rule intended for predicting a feature (high CN) that is almost exclusive of residues placed in the core of a protein.

These two observations illustrate how easy is to interpret the solutions generated by GAssist, in opposition to decision trees of 8,000 leaves, such as the ones that C4.5 can produce on these datasets, or the almost inexistent explanatory power of LIBSVM or Naive Bayes. GAssist can provide added value to the PSP experts, as not only has pretty good data mining capacity, but also it can do knowledge discovery via the explanatory power of the solutions it produces.

Table 12 extends this analysis to all the rule-sets generated for this dataset, reporting two metrics: (a) the frequency of appearance of each AA type for each window position, and (b) the average appearance frequency of each AA type for all positions. From this average we obtain a ranking of specificity of each AA type: Glutamine (E) and Proline (P) are the two AA types appearing less often. On the other hand Alanine (A), Cysteine (C), Phenylalanine (F), Isoleucine (I), Leucine (L), Methionine (M) and Valine (V) appear in more than 95% of all positions, therefore being the less specific AA types for predicting a high value for the CN. Also, all of these latter residues are hydrophobic.

Table 13 analyzes these rules from a slightly different point of view: ranking the AA types for their frequency of appearance in each window position. Previously we observed that the predicate for the central residue of all rules takes a different form compared to the rest of predicates. This issue is reflected perfectly by this ranking. For the central residue, after the 8th AA type in the ranking all frequencies are very close to 0, for the other ones, we do not find a frequency less than 95% until position 15th of the ranking. For the non-central positions, the interesting columns are the ones at the bottom of the ranking. We can observe that Proline (P) and Glutamine (E) are the less frequent AA types for seven of the nine window positions.

Therefore, we can extract sound explanations from the generated rules, and we have found more paths of analysis: analyzing the specificity degree of the used attributes and window positions, and relating the predicates generated by GAssist with physical/chemical properties.

6 Discussion

In this section we will discuss the results presented in the previous section. As the rest of the chapter, this discussion is divided in two main parts, corresponding to the two reported batteries of experiments. Moreover, we will also briefly describe some other research work we have done which is related to this chapter.

6.1 Battery of Experiments I

The LCS and other machine learning algorithms preformed at similar levels for these CN prediction tasks. Generally, increasing the number of states leads to a reduction in prediction accuracy. Reduction of input information from full residue type to HP sequence reduces the accuracy of prediction. The algorithms were, however, all capable of predictions using HP sequence that were within 4% of the accuracies obtained using full residue type sequences, considering that the size of the representation is ten times smaller (20-letter alphabet vs. 2-letter alphabet).

Table 12. Frequency of appearance in percentage of each AA type by window position in the generated rules for the CN1 dataset and uniform-length class definitions

Pos.	A	C	D	E	F	G	H	I	K	L	M	N	P	Q	R	S	T	V	W	Y
−4	100	96	100	44	99	100	98	100	93	100	100	100	100	84	94	100	100	100	83	99
−3	100	94	14	12	100	100	100	100	100	100	100	89	94	66	100	100	100	100	94	99
−2	100	94	100	98	100	100	99	100	100	100	99	97	88	98	99	98	99	100	94	100
−1	100	94	41	43	98	100	100	100	18	100	100	100	8	96	99	100	100	100	96	100
0	100	100	0	0	100	0	0	100	0	100	100	0	0	0	0	0	0	100	82	5
1	100	97	4	4	100	99	98	100	96	100	100	81	0	94	100	99	100	100	99	100
2	100	98	98	100	100	100	98	100	97	100	97	100	38	100	100	100	100	100	98	100
3	100	98	55	11	100	100	100	100	1	100	100	96	2	47	66	100	99	100	100	100
4	100	99	94	1	100	100	98	100	1	100	100	96	66	1	28	100	100	100	85	100
Ave.	100.0	96.7	56.2	34.8	99.7	88.8	87.9	100.0	56.2	100.0	99.6	84.3	44.0	65.1	76.2	88.6	88.7	100.0	92.3	44.6

Table 13. Ranking of appearance of the AA type by window position in the generated rules for the CN 1 dataset and uniform-length class definitions

Pos.	1	2	3	4	5	6	7	8	9	10	11	12	13	14	15	16	17	18	19	20
−4	V-100	T-n100	S-100	P-100	N-100	M-100	L-100	I-100	G-100	D-100	A-100	Y-99	F-99	H-98	C-96	R-94	K-93	Q-84	W-83	E-44
−3	V-100	T-100	S-100	R-100	M-100	L-100	K-100	I-100	H-100	G-100	F-100	A-100	Y-99	W-94	P-94	C-94	N-89	Q-66	D-14	E-12
−2	Y-100	V-100	L-100	K-100	I-100	G-100	F-100	D-100	A-100	T-99	R-99	M-99	H-99	S-98	Q-98	E-98	N-97	W-94	C-94	P-88
−1	Y-100	V-100	T-100	S-100	N-100	M-100	L-100	I-100	H-100	G-100	A-100	R-99	F-98	W-96	Q-96	C-94	E-43	D-41	K-18	P-8
0	V-100	M-100	L-100	I-100	F-100	C-100	A-100	W-82	Y-5	T-0	S-0	R-0	Q-0	P-0	N-0	K-0	H-0	G-0	E-0	D-0
1	Y-100	V-100	T-100	R-100	M-100	L-100	I-100	F-100	A-100	W-99	S-99	G-99	H-98	C-97	K-96	Q-94	N-81	E-4	D-4	P-0
2	Y-100	V-100	T-100	S-100	R-100	Q-100	N-100	L-100	I-100	G-100	F-100	E-100	A-100	W-98	H-98	D-98	C-98	M-97	K-97	P-38
3	Y-100	W-100	V-100	S-100	M-100	L-100	I-100	H-100	G-100	F-100	A-100	T-99	C-98	N-96	R-66	D-55	Q-47	E-11	P-2	K-1
4	Y-100	V-100	T-100	S-100	M-100	L-100	I-100	G-100	F-100	A-100	C-99	H-98	N-96	D-94	W-85	P-66	R-28	Q-1	K-1	E-1

For all of the algorithms studied, in the case of the most informative five state predictions, moving from HP lattice to real protein HP sequences leads to a reduction of CN prediction accuracy from levels of around 50% to levels of around 30%. The significant reduction in the spatial degrees of freedom in the Lattice-HP models leads to an improvement in prediction accuracy of around 20%.

In contrast, moving from the real protein HP sequences to real protein full residue type sequences (for the same five state CN predictions) only a 3–5% improvement in prediction accuracy results from inclusion of this additional residue type information. This observation matches the general agreement in the computation biology community that hydrophobicity is one of the main properties that guides the folding process of proteins and, thus, it is a key determinant of good CN prediction, and also that algorithmic studies of HP models are relevant.

The rules that result from a reduced two letter alphabet are simpler and easier to understand than those from the full residue type studies. For example, for the HP representation a rule set giving 62.9% accuracy is shown below (an X symbol is used to represent positions where the sliding window overlaps with the end of the chain).

1. If $AA_{-1} \notin \{x\}$ and $AA \in \{h\}$ and $AA_1 \in \{p\}$ then class is 1
2. If $AA_{-1} \in \{h\}$ and $AA \in \{h\}$ and $AA_1 \notin \{x\}$ then class is 1
3. If $AA_{-1} \in \{p\}$ and $AA \in \{h\}$ and $AA_1 \in \{h\}$ then class is 1
4. Default class is 0

In these rules, a class assignment of high is represented by 1 and low by 0. For the full residue type representation a rule set giving 67.7% accuracy is:

1. If $AA_{-1} \notin \{D, E, K, N, P, Q, R, S, X\}$ and $AA \notin \{D, E, K, N, P, Q, R, S, T\}$ and $AA_1 \notin \{D, E, K, Q, X\}$ then class is 1
2. If $AA_{-1} \notin \{X\}$ and $AA \in \{A, C, F, I, L, M, V, W, Y\}$ and $AA_1 \notin \{D, E, H, Q, S, X\}$ then class is 1
3. If $AA_{-1} \notin \{P, X, Y\}$ and $AA \in \{A, C, F, I, L, M, V, W, Y\}$ and $AA_1 \notin \{K, M, T, W, X, Y\}$ then class is 1
4. If $AA_{-1} \notin \{H, I, K, M, X\}$ and $AA \in \{C, F, I, L, M, V, W, Y\}$ and $AA_1 \notin \{M, X\}$ then class is 1
5. Default class is 0

6.2 Battery of Experiments II

The discussion for this battery of experiments is centered on the comparison of learning methods in these datasets. GAssist performs better than Naive Bayes and C4.5 but worse than LIBSVM. The direct comparison of GAssist and C4.5 is clearly favorable to GAssist, and this is important as these two systems use a very similar knowledge representation (axis-parallel). GAssist is better than C4.5 at exploring the search space of the solutions that this kind of knowledge representation can offer. On the other hand, LIBSVM managed to outperform GAssist in most datasets, especially in those with real-valued input attributes, which may indicate that the non-linear knowledge representation used by LIBSVM is superior for this kind of data.

Nevertheless, there are several strong points that back the use of GAssist. The first of them, analyzed in the results section, is the explanatory power of the solutions that GAssist generates. From a pure machine learning point of view these solutions are extremely compact, both in number of rules and also in number of expressed attributes. Moreover, as we have shown, it is quite easy to extract practical real-world explanations from the generated rules. On the other hand, it is quite difficult to extract an explanation from LIBSVM solutions. The only complexity measure that LIBSVM provides about its solutions is the number of instances selected as support vectors, and this number can be huge, around 70–90% of the training set.

Another issue of concern is the run-time. Although GAssist is not a fast learning system, it is considerably faster than LIBSVM. GAssist run time on these datasets ranged between 0.3 and 24 h, while LIBSVM run time ranged from 21 h to 4.5 days. Even more critical is the time spent at the test stage. While LIBSVM in some cases took hours to predict all examples in the test set (mainly because of the high number of support vectors as stated above), GAssist used approximately about a minute to use its ensemble of rule-sets to produce the test predictions.

This issue is very important, because the final goal of the line of research where this work is included is to create an on-line web-based 3D protein structure prediction server, which integrates the coordination number predictors and also other related PSP datasets such as secondary structure prediction, solvent accessibility and disulfide bonding. Such a server would be queried simultaneously by multiple users that would normally want to predict tens, if not hundreds, of protein structural features with as few time delays as possible. Our experience with the www.proksi.net web server for protein structure comparison indicates that in an exploitation environment such as this the run-time is critical, and in this aspect GAssist can be faster than LIBSVM by two orders of magnitude.

6.3 Brief Description of Other Related Work

The work presented in this chapter is just a part of an outgoing line of research funded by the UK Engineering and Physical Sciences Research Council. We have continued our efforts in this research line and in this subsection we would like to briefly summarize some of the most closely related of them to the experiments described in this chapter.

The first of them [47] studies an automated method to produce alphabet reduction of the primary sequence. As we have seen in this chapter, there is a certain performance gap (less than 4%) between the accuracy obtained using the HP representation and the accuracy obtained using the AA representation. Can this gap be significantly reduced using a slightly higher alphabet size than two? We used an automated information theory-based evolutionary computation method to find the proper alphabet reduction policy, and tailor it specifically for the PSP feature being predicted. The method uses the Extended Compact Genetic Algorithm (ECGA) [48] using the Mutual Information metric [49] as fitness function. Afterwards, the produced alphabet reduction was verified by learning the dataset with reduced alphabet using BioHEL [50], a recent Learning Classifier System using the Iterative Rule Learning paradigm [51] combined with several of the features of GAssist such as the MDL fitness function, the explicit default rule mechanism and the ILAS windowing scheme. Our experiments determined that we can produce reduced alphabets applied to the Coordination Number prediction dataset with only three letters that can obtain an accuracy which is only 0.6% lower than the accuracy obtained by the AA representation. Therefore, it represents substantial progres when compared to the standard reduced alphabets used in the literature such as the HP alphabet.

Another related work [52] studies alternative definitions of the coordination number metric based on alternative ways of defining the neighbourhood (the residues in contact with) of a certain residue. These neighbourhood definitions are based on graph theory, specifically on Proximity Graphs [53], such as the Delaunay Tessellation and the Minimum Spanning Tree, among others. Four neighbourhood definitions that produced four alternative coordination number metrics were tested, using all the protocol of class partitions criteria and the six sets of input data described in the second battery of experiments of this chapter. The evaluation process identified which measures are easier to predict than others. The explanatory power of the produced rules was also analyzed.

7 Conclusions and Further Work

In this chapter we have described our recent experiments with Learning Classifier Systems applied to a Bioinformatics problem called Coordination Number prediction. This problem belongs to the family of problems derived from Protein Structure Prediction. The associated datasets are a challenge to LCS for many reasons: (a) very large datasets with at least hundreds of thousands of instances, (b) noisy data and (c) challenging feature selection due to the incomplete expert understanding of these domains.

We applied a Pittsburgh approach LCS called GAssist to several variants of this problem and compared its performance to some standard machine learning methods. In general GAssist showed good performance, only being outperformed by Support Vector Machines in some of the datasets. Nevertheless, GAssist showed a competitive advantage against SVM in some aspects, especially: run-time in an exploitation environment and explanatory power.

It is very difficult to understand the rationale behind SVM predictions. On the other hand, GAssist produces very small and compact rule sets for all datasets. These rule sets were easy to interpret by the domain experts and the explanations behind the predictions were acknowledged to be sound.

We have assessed two classes of input information about this domain, how to represent the primary sequence of the protein and what information we can add to improve the CN prediction, as well as assessing several definitions of classes for the coordination number prediction domain. The performance difference between the HP and the AA representations is significant but not huge. In future work we would like to investigate if we can find other kinds of reduced alphabets where this performance difference becomes minimal. We have already started to obtain successful results in this area, as briefly summarized in the discussion section. We would also like to evaluate other kinds of input information for CN prediction not tested yet, such as position specific scoring matrices or relative solvent accessibility, as well as assessing what is the real quantity of information that each of the tested class definitions is able to provide, as the final goal of predicting CN is to integrate this predictions into a 3D protein structure prediction system. The explanatory analysis of the generated rule-sets would also be very useful, and not just to understand the GAssist predictions, but also in order to identify information that can be fed back to GAssist to improve its learning process and to better understand aspects of protein folding. Finally, we would like to investigate how can we improve the learning process of GAssist, in order to improve its performance and scalability. We have also started to obtain better results in this scope with the development of BioHEL [50], which has better scalability capacity than GAssist, and manages to obtain similar performance to LIBSVM in many datasets. Other interesting alternatives could be hyper-ellipsoidal conditions [54], neural network conditions [55] or some kind of SVM-LCS hybrid.

Acknowledgments

We acknowledge the support of the UK Engineering and Physical Sciences Research Council (EPSRC) under grants GR/T07534/01, EP/D061571/1 and GR/S64530/01 and the Biotechnology and Biological Sciences Research Council (BBSRC) under grant BB/C511764/1. We are grateful for the use of the University of Nottingham's High Performance Computer.

References

1. Holland, J.H., Reitman, J.S.: Cognitive systems based on adaptive algorithms. In Hayes-Roth, D., Waterman, F., eds.: Pattern-Directed Inference Systems. Academic, New York (1978) 313–329

2. Smith, S.: A Learning System Based on Genetic Algorithms. PhD thesis, University of Pittsburgh, Pittsburgh (1980)
3. Bernadó, E., Llorà, X., Garrell, J.M.: XCS and GALE: a comparative study of two learning classifier systems with six other learning algorithms on classification tasks. In: Fourth International Workshop on Learning Classifier Systems – IWLCS-2001. (2001) 337–341
4. Bacardit, J., Butz, M.V.: Data mining in learning classifier systems: comparing xcs with gassist. In: Advances at the frontier of Learning Classifier Systems. Springer, Berlin Heidelberg New York (2007) 282–290
5. Wilson, S.W.: Classifier fitness based on accuracy. Evolutionary Computation **3** (1995) 149–175
6. Llorà, X., Garrell, J.M.: Knowledge-independent data mining with fine-grained parallel evolutionary algorithms. In: Proceedings of the Third Genetic and Evolutionary Computation Conference. Morgan Kaufmann, San Francisco (2001) 461–468
7. Bacardit, J.: Pittsburgh Genetics-Based Machine Learning in the Data Mining Era: Representations, Generalization, and Run-Time. PhD thesis, Ramon Llull University, Barcelona, Spain (2004)
8. Stout, M., Bacardit, J., Hirst, J.D., Krasnogor, N., Blazewicz, J.: From HP lattice models to real proteins: coordination number prediction using learning classifier systems. In: Applications of Evolutionary Computing, EvoWorkshops 2006, Springer, Berlin Heidelberg New York, LNCS 3907 (2006) 208–220
9. Bacardit, J., Stout, M., Krasnogor, N., Hirst, J.D., Blazewicz, J.: Coordination number prediction using learning classifier systems: performance and interpretability. In: GECCO'06: Proceedings of the 8th Annual Conference on Genetic and Evolutionary Computation. ACM Press, New York (2006) 247–254
10. Stout, M., Bacardit, J., Hirst, J.D., Krasnogor, N.: Prediction of residue exposure and contact number for simplified hp lattice model proteins using learning classifier systems. In Ruan, D., D'hondt, P., Fantoni, P.F., Cock, M.D., Nachtegael, M., Kerre, E.E., eds.: Proceedings of the 7th International FLINS Conference on Applied Artificial Intelligence. World Scientific, Genova (2006) 601–608
11. Hinds, D.A., Levitt, M.: A lattice model for protein-structure prediction at low resolution. Proceedings of the National Academy Sciences of the United States of America **89** (1992) 2536–2540
12. Yue, K., Fiebig, K.M., Thomas, P.D., Sun, C.H., Shakhnovich, E.I., Dill, K.A.: A test of lattice protein folding algorithms. Proceedings of the National Academy Sciences of the United States of America **92** (1995) 325–329
13. Kinjo, A.R., Horimoto, K., Nishikawa, K.: Predicting absolute contact numbers of native protein structure from amino acid sequence. Proteins **58** (2005) 158–165
14. Baldi, P., Pollastri, G.: The principled design of large-scale recursive neural network architectures dag-rnns and the protein structure prediction problem. Journal of Machine Learning Research **4** (2003) 575–602
15. Shao, Y., Bystroff, C.: Predicting interresidue contacts using templates and pathways. Proteins **53** (2003) 497–502
16. MacCallum, R.: Striped sheets and protein contact prediction. Bioinformatics **20** (2004) I224–I231

17. Zhao, Y., Karypis, G.: Prediction of contact maps using support vector machines. In: Proceedings of the IEEE Symposium on BioInformatics and BioEngineering (2003) 26–36
18. Altschul, S.F., Madden, T.L., Scher, A.A., Zhang, J., Zhang, Z., Miller, W., Lipman, D.J.: Gapped blast and psi-blast: a new generation of protein database search programs. Nucleic Acids Research **25** (1997) 3389–3402
19. Abe, H., Go, N.: Noninteracting local-structure model of folding and unfolding transition in globular proteins. Part 2. Application to two-dimensional lattice proteins. Biopolymers **20** (1981) 1013–1031
20. Hart, W.E., Istrail, S.: Crystallographical universal approximability: a complexity theory of protein folding algorithms on crystal lattices. Technical Report SAND95-1294, Sandia National Labs, Albuquerque (1995)
21. Hart, W., Istrail, S.: Robust proofs of NP-hardness for protein folding: general lattices and energy potentials. Journal of Computational Biology (1997) 1–20
22. Escuela, G., Ochoa, G., Krasnogor, N.: Evolving l-systems to capture protein structure native conformations. In: Proceedings of the 8th European Conference on Genetic Programming (EuroGP 2005), Lecture Notes in Computer Sciences 3447, pp. 73–84, Springer, Berlin Heidelberg New York (2005)
23. Krasnogor, N., Pelta, D.: Fuzzy memes in multimeme algorithms: a fuzzy-evolutionary hybrid. In Verdegay, J., ed.: Fuzzy Sets based Heuristics for Optimization. Springer, Berlin Heidelberg New York (2002)
24. Krasnogor, N., Hart, W., Smith, J., Pelta, D.: Protein structure prediction with evolutionary algorithms. In Banzhaf, W., Daida, J., Eiben, A., Garzon, M., Honavar, V., Jakaiela, M., Smith, R., eds.: GECCO-99: Proceedings of the Genetic and Evolutionary Computation Conference. Morgan Kaufmann (1999)
25. Krasnogor, N., Blackburne, B., Burke, E., Hirst, J.: Multimeme algorithms for protein structure prediction. In: Proceedings of the Parallel Problem Solving from Nature VII. Lecture Notes in Computer Science. Volume 2439 (2002) 769–778
26. Krasnogor, N., de la Cananl, E., Pelta, D., Marcos, D., Risi, W.: Encoding and crossover mismatch in a molecular design problem. In Bentley, P., ed.: AID98: Proceedings of the Workshop on Artificial Intelligence in Design 1998 (1998)
27. Krasnogor, N., Pelta, D., Marcos, D., Risi, W.: Protein structure prediction as a complex adaptive system. In: Proceedings of Frontiers in Evolutionary Algorithms 1998 (1998)
28. DeJong, K.A., Spears, W.M., Gordon, D.F.: Using genetic algorithms for concept learning. Machine Learning **13** (1993) 161–188
29. Bacardit, J.: Analysis of the initialization stage of a pittsburgh approach learning classifier system. In: GECCO 2005: Proceedings of the Genetic and Evolutionary Computation Conference. Volume 2., ACM Press, New York (2005) 1843–1850
30. Rissanen, J.: Modeling by shortest data description. Automatica **14** (1978) 465–471
31. Bacardit, J., Goldberg, D., Butz, M., Llorà, X., Garrell, J.M.: Speeding-up pittsburgh learning classifier systems: Modeling time and accuracy. In: Parallel Problem Solving from Nature - PPSN 2004, Springer, Berlin Heidelberg New York, LNCS 3242 (2004) 1021–1031
32. Breiman, L.: Bagging predictors. Machine Learning **24** (1996) 123–140
33. Bacardit, J., Krasnogor, N.: Empirical evaluation of ensemble techniques for a pittsburgh learning classifier system. In: Proceedings of the 9th International Workshop on Learning Classifier Systems. (to appear), LNAI, Springer (2007)

34. Blake, C., Keogh, E., Merz, C.: UCI repository of machine learning databases (1998) (www.ics.uci.edu/mlearn/MLRepository.html)
35. Liu, H., Hussain, F., Tam, C.L., Dash, M.: Discretization: an enabling technique. Data Mining and Knowledge Discovery **6** (2002) 393–423
36. Noguchi, T., Matsuda, H., Akiyama, Y.: Pdb-reprdb: a database of representative protein chains from the protein data bank (pdb). Nucleic Acids Research **29** (2001) 219–220
37. Sander, C., Schneider, R.: Database of homology-derived protein structures. Proteins **9** (1991) 56–68
38. Broome, B., Hecht, M.: Nature disfavors sequences of alternating polar and non-polar amino acids: implications for amyloidogenesis. Journal of Molecular Biology **296** (2000) 961–968
39. Quinlan, J.R.: C4.5: Programs for Machine Learning. Morgan Kaufmann (1993)
40. John, G.H., Langley, P.: Estimating continuous distributions in Bayesian classifiers. In: Proceedings of the Eleventh Conference on Uncertainty in Artificial Intelligence. Morgan Kaufmann Publishers, San Mateo (1995) 338–345
41. Witten, I.H., Frank, E.: Data Mining: Practical Machine Learning Tools and Techniques with Java Implementations. Morgan Kaufmann (2000)
42. Miller, R.G.: Simultaneous Statistical Inference. Springer, Berlin Heidelberg New York (1981)
43. Jones, D.: Protein secondary structure prediction based on position-specific scoring matrices. Journal of Molecular Biology **292** (1999) 195–202
44. Chang, C.C., Lin, C.J.: LIBSVM: A library for support vector machines. Department of Computer Science and Information Engineering, National Taiwan University. (2001) Software available at http://www.csie.ntu.edu.tw/~cjlin/libsvm.
45. Booker, L.: Recombination distribution for genetic algorithms. In: Foundations of Genetic Algorithms 2. Morgan Kaufmann (1993) 29–44
46. Livingstone, C.D., Barton, G.J.: Protein sequence alignments: a strategy for the hierarchical analysis of residue conservation. Computer Applications in the Biosciences **9** (1993) 745–756
47. Bacardit, J., Stout, M., Hirst, J.D., Sastry, K., Llorà, X., Krasnogor, N.: Automated alphabet reduction method with evolutionary algorithms for protein structure prediction. In: GECCO'07: Proceedings of the 9th Annual Conference on Genetic and Evolutionary Computation. ACM Press, New York (2007) to appear
48. Harik, G.: Linkage learning via probabilistic modeling in the ecga. Technical Report 99010, Illinois Genetic Algorithms Lab, University of Illinois at Urbana-Champaign (1999)
49. Cover, T.M., Thomas, J.A.: Elements of Information Theory. Wiley, New York (1991)
50. Bacardit, J., Krasnogor, N.: Biohel: bioinformatics-oriented hierarchical evolutionary learning. Nottingham eprints, University of Nottingham (2006)
51. Venturini, G.: Sia: a supervised inductive algorithm with genetic search for learning attributes based concepts. In Brazdil, P.B., ed.: Machine Learning: ECML-93 - Proceedings of the European Conference on Machine Learning. Springer, Berlin Heidelberg New York (1993) 280–296
52. Stout, M., Bacardit, J., Hirst, J.D., Smith, R.E., Krasnogor, N.: Prediction of topological contacts in proteins using learning classifier systems. Soft Computing

(2007) Special Issue on Evolutionary and Metaheuristic–based Data Mining (EMBDM), to appear
53. Preparata, F.P.: Computational geometry : an introduction/Franco P. Preparata, Michael Ian Shamos. Texts and monographs in computer science. Springer (1985)
54. Butz, M.V., Lanzi, P.L., Wilson, S.W.: Hyper-ellipsoidal conditions in xcs: rotation, linear approximation, and solution structure. In: GECCO'06: Proceedings of the 8th annual conference on Genetic and evolutionary computation. ACM Press, New York (2006) 1457–1464
55. O'Hara, T., Bull, L.: Backpropagation in accuracy-based neural learning classifier systems. In: Advances at the frontier of Learning Classifier Systems. Springer, Berlin Heidelberg New York (2007) 25–39

Improving Evolutionary Computation Based Data-Mining for the Process Industry: The Importance of Abstraction

William N.L. Browne

Department of Cybernetics, University of Reading, Whiteknights, Reading, Berkshire RG6 6AY, UK, w.n.browne@reading.ac.uk

Summary. An Industrial Learning Classifier System (LCS) was developed a decade ago for the mining of information in process industries, specifically for a Steel Hot Strip Mill. Despite encouraging results, the prediction accuracy achieved and the effort required did not warrant adoption. The lessons learnt for applying Genetic-based Machine Learning to industrial data-mining applications are still relevant and are described here. After 10 years further research into LCS much innovation has occurred: messy encodings, rule-base reduction, pass-through rules and flexible encodings. However, this paper hypothesises that the biggest hurdle preventing LCS from wider industrial adoption is a lack of 'abstraction', i.e., after states have been linked to actions and generalised by removing irrelevant information, similarities between rules must be abstracted to form higher level rules. Initial results for a 'toy' problem demonstrate that LCS are capable of abstraction with powerful consequences.

1 Introduction

Industrial domains seek to maximise profits from existing plant due to the large capital costs and long-term payback times of new plant. Modern data-capture technology is comparatively low cost and produces a data resource. This collected data has the potential to give prediction and diagnosis of plant input (e.g. set up) to output (e.g. product quality) from the information contained. Many bespoke and commercial data-mining tools exist, but the Genetics based Machine Learning technique of Learning Classifier Systems (LCS) has unique properties that could give commercial advantage if developed for such industrial domains.

LCS build a population of transparent rules that co-operate to map environmental input to useful outputs. There is no one LCS algorithm, but many complementary (or conflicting) methods that can be employed in the LCS concept. This allows LCS the flexibility to solve problems in multiple domains, such as stimulus-response (e.g. data-mining), ultimate-response

W.N.L. Browne: *Improving Evolutionary Computation Based Data-Mining for the Process Industry: The Importance of Abstraction*, Studies in Computational Intelligence (SCI) **125**, 47–68 (2008)
www.springerlink.com

(e.g. bin-packing) and continuous response (e.g. robot navigation). Different families of LCS have evolved from the original concept by John Holland [1]. However, it is not obvious which LCS are best suited to a given domain and what methods to utilise within them.

The data-mining project described was from 1994 to 1998, which has meant a decade of subsequent development in the LCS technique. Many of the criticisms of the tried and tested techniques available then have been addressed subsequently. This has been necessary as the complexity of the steel mill data overwhelmed the capabilities of the pre-1994 LCS. Instead of waiting for better methods to be evolved (hindsight showed this occurred quickly with the introduction of niche based rule discovery and accuracy of prediction in the ZCS and XCS techniques [2, 3]), the industrial environment was examined to form the basis of an industrialised LCS technique. This unique starting point lead to insight into the operation of the LCS technique, novel methods being developed and useful information being extracted from complex industrial data.

This chapter seeks to show how analysis of the application domain can lead to the selection of the LCS platform and methods for further development. Novel methods and a complete industrialised LCS (termed iLCS) platform are presented with supporting results. A major issue for this (and many other) industrial data-mining techniques was identified in that similarities between important patterns were not abstracted to higher-level rules. Therefore, a novel abstraction algorithm is presented, see Fig. 1, and tested on a toy problem, to address this deficiency in LCS for data mining.

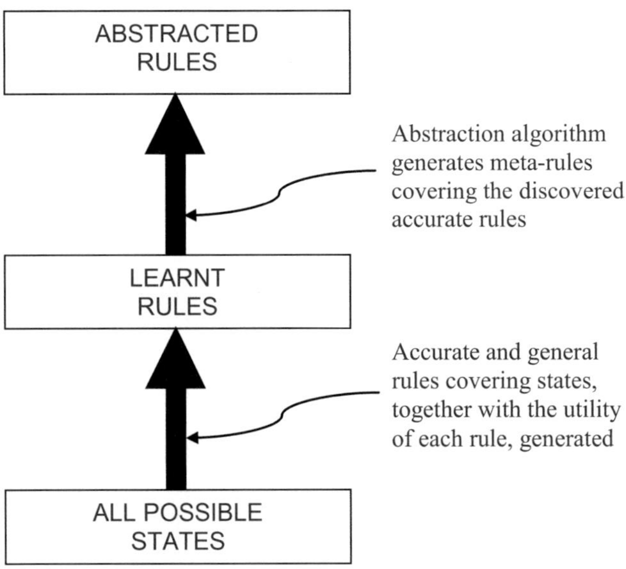

Fig. 1. Schematic of abstraction to produce higher-order rules

The chapter starts with the background to the application domain, with the learning considerations and comparison with basic LCS available at the time. The task description gives the aims and objectives of the project which helps to focus the development. The development and novel methods of the industrial LCS are outlined followed by supporting results and discussion. Abstraction is motivated, described and initial results presented. Conclusions are drawn based on the industrial data-mining and abstraction experiments.

2 Background

2.1 Description of the HSM and Downcoilers of Llanwern Works

Figure 2 shows the general layout of the Hot Strip Mill (HSM) in Llanwern British Steel Strip Products (now Corus) works where the majority of the data has been taken. Steel slabs of approximately $10\,\mathrm{m} \times 2\,\mathrm{m} \times 0.4\,\mathrm{m}$ and 20 tonnes, are re-heated in furnaces to a critical rolling temperature. The finishing stands further reduce the thickness and, together with the water sprays, induce the desired material properties.

The head end of strip is guided into the downcoiler by hydraulic side guides. Incorrect entry may lead to a 'cobble', where the strip jams in the downcoiler, or a pinch, where the strip folds over on itself, and results in loss of material quality or scrap.

The data mining task was to link Mill setup parameters to product quality ratings in order to predict and diagnose mill quality problems.

2.2 LCS for Data Mining

Up to 250 mill parameters could be input into the LCS, but training for such a LCS would be lengthy with the results produced complex and non-transparent. Both domain knowledge and statistics must be used to reduce

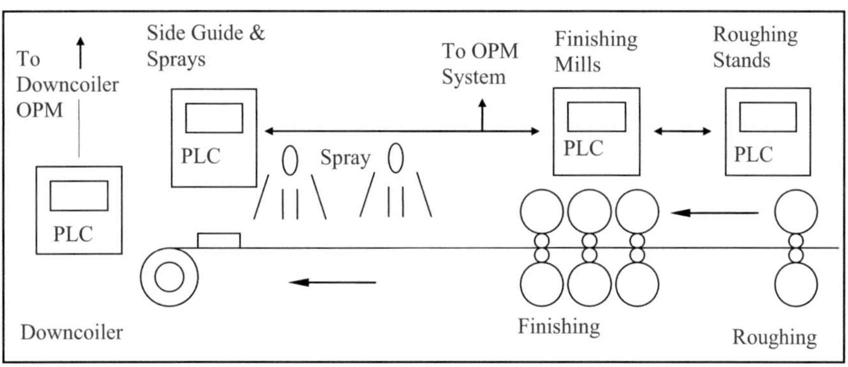

Fig. 2. Schematic showing a downcoiler and programmable logic controllers

the dimensionality of the data, but care that bias and assumptions do not remove important/vital pieces of information must be taken.

Noise, contradiction within data, missing data, accuracy and consistency of readings and epistasis (where the value of one parameter affects the importance of the value of another) all contribute to a complex problem domain, see Booker [4]. LCS are suited to the proposed environment because they can learn knowledge from imperfect data taken from real environments Wilson [5].

The state of LCS technique prior to the start of the project was best summarised Goldberg [6]: "LCS are a quagmire – a glorious, wondrous and inventing quagmire, but a quagmire nonetheless". The quagmire exists because of the development of computers and the consequent potential for machine learning paradigms – not because the LCS technique has fundamental flaws.

3 Development of iLCS

The iLCS was developed as a whole concept, see Fig. 3. Therefore, care must be taken if adopting individual developments, as they may not function as described without the other supporting methods. A comprehensive description of iLCS is presented in Bull [7] with the main features summarised here.

The balance needed between exploring and exploiting (E/E) information is a critical problem for industrial domains as the optimal balance level depends on both the domain and the stage of training as more knowledge is progressively gained. Too much exploration results in the global optimum never being reached, whilst too much exploitation traps the system in local optima.

The internal parameters of the LCS were kept simple, flexible and robust to reduce the time-consuming effort of setting evolutionary computation parameters. A Michigan style LCS was adopted due to its flexibility to adjust to unknown structures in environments (If the rule structure is known or thought to be simple, then a Pittsburgh approach may be preferable).

Considering that the desirable E/E balance changes depending on the stage of training and the environment, an LCS that has a static balance, which is hard to adjust, may not be suited to the complex domains found in industry. Simulated Annealing type schemes, which reduce the amount of exploration as training progresses, have been tested [8], but these are hard to control.

Investigating the stages of training in simple domains showed three main phases: exploring for information, combining exploring and exploiting to develop useful rules and then stabilising the rules into exploitable hierarchies. Therefore, the rule-base was split into three distinct phases, linked by a stepping-stone approach to transfer rules from search to combine to stable phases. Long surviving rules could also be fed back to the search population to guide search into areas of interest. Different methods that support the purpose of each phase were utilised, e.g. roulette, tournament and then rank selection.

The fitness of a classifier is split into two measures: *Fertility* controls selection in rule discovery and *Effectiveness* controls selection in the auction.

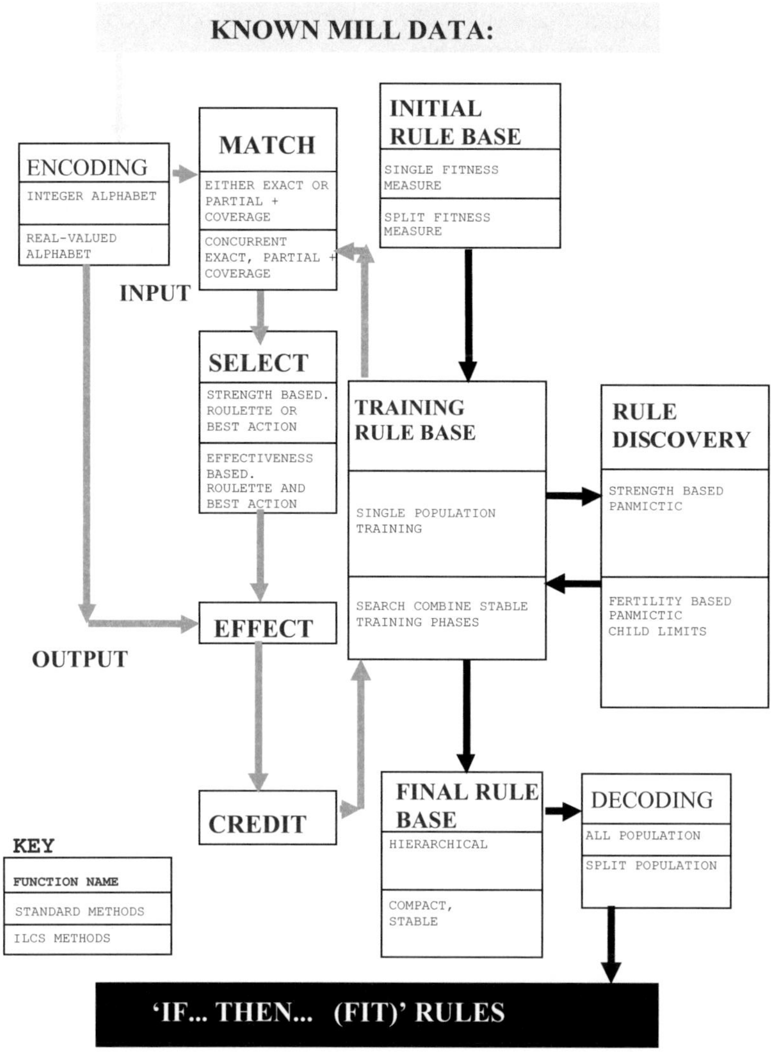

Fig. 3. The concept of iLCS (*grey shading*) compared with basic LCS (*white*)

Fertility exerts a generalisation pressure as it combines a measure of a classifier's generality with its accuracy. Effectiveness is the expected environmental pay-off (similar to strength based systems), but could easily be changed to other measures (e.g. accuracy).

The advances and novel aspects in iLCS were:

- Real alphabet – encoding matched domain.
- Phases of training – methods used at appropriate times reduced parameter tuning.

- Evaluation limits – new rules accommodated into population, whilst child limit made stagnation less likely.
- Panmictic Rule discovery – complete search in initial phase.
- Split fitness measure – generalisation pressure focused.
- Removal of taxation and genotypic specificity measures for mating – simplified operation.

4 Initial Data-Mining Results

The data from the mill contains information requiring discovery, but the quality of this information is unknown. If the information learnt by a LCS that has only been operated on real data is poor, then it is impossible to determine whether it is poor operation by the LCS on good data or good operation on poor data. Confidence can be gained from analysis of the rules produced and previous good results with simulated data, prior to testing with real data. The simulated data had eight conditions linked to one action that could take four different values of different base rate. Results, see Fig. 4 and Table 1, encourage the use with real data as 100% prediction is reached with iLCS identifying very small niches.

The benefits of an industrial LCS have been split into two categories – Discovery and Advisory – which depend upon the use of the LCS. Discovery is classified as the LCS working off-line on off-line data in an attempt to gain plant knowledge (diagnosis of plant operation). The Advisory stage takes the evaluated knowledge and provides off-line predictions from on-line data (prediction of plant operation). The predictions are interpreted by human operators to decide if any action is required. This project was not designed to provide on-line actions from on-line information as the consequences of

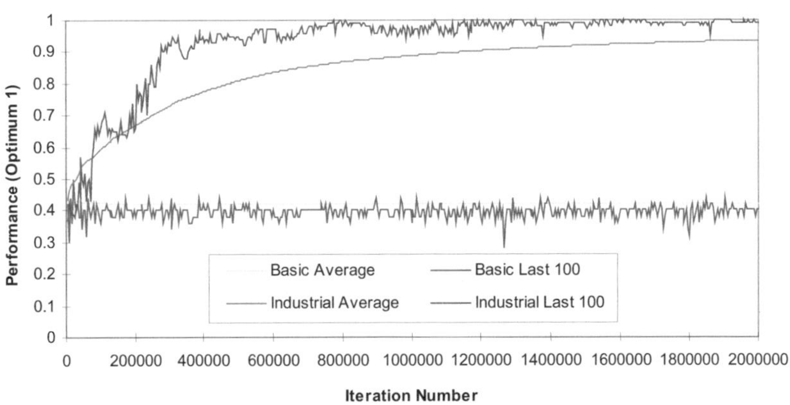

Fig. 4. Training performance of Industrial LCS compared with Basic LCS

Table 1. Results from simulated data

Feature extraction classifier										
Rank	Parameter - upper limit, lower limit and resolution								Out	Specificity
1	101.7	1,601.4	5.5	236.5	1,600.5	1,000.5	750.5	3.5	[0]	27
	0	0	1.5	200.0	1,000.0	800.0	550.2	0.5		
	10	96	1	5.1	60	20	20	0.8		
4	150.5	2,252.8	5.5	250.5	1,600.5	980.5	750.5	3.5	[1]	16
	0	0	1.5	240.4	1,012.0	800.0	550.0	0.5		
	10	96	1	5.1	60	20	20	0.8		
2	150.5	2,120.0	5.5	244.4	1,600.5	1,000.5	750.5	3.5	[2]	24
	101.5	0	1.5	200.0	1,000.1	800.0	550.0	0.5		
	10	96	1	5.1	60	20	20	0.8		
3	101.7	2,141.0	5.5	250.5	1,600.5	1,000.5	750.5	3.5	[3]	20
	0	0	1.5	200.0	1,000.1	825.0	550.0	1.3		
	10	96	1	5.1	60	20	20	0.8		
5	100.7	2,141.0	5.5	240.5	1,600.5	1,000.5	750.5	3.5	[0]	22
	0	1,600.0	1.5	200.0	1,000.1	800	550.0	1.3		
	10	96	1	5.1	60	20	20	0.8		

incorrect action were considered too serious to risk using a novel learning process.

4.1 Discovery Training Performance

The long time taken to train was a major disadvantage of the LCS technique in industrial domains. Modern computing power has reduced discovery training to a few hours (note advisory performance is near instantaneous from a learnt rule-base).

Adjusting the representation of the domain in the training set would also improve times.

- Reduced number of parameters in the training set
- Sufficient number of training examples to learn all quality levels
- Introduction of correct and non-biasing domain knowledge.

It is recognised that improved efficiency and effectiveness of the LCS can reduce training times.

The shape of the performance graphs for the industrial LCS with industrial data, show that learning of useful information did occur (e.g. Fig. 5). Performance on all industrial data sets was better than a 'random' (pick any output) or 'greedy' (pick the most common output) tactic.

Despite learning useful information, the LCS did not obtain satisfactory performance. Regular peaks and troughs of performance indicate that the LCS was continually searching for better information, which necessarily included rejecting previously useful information.

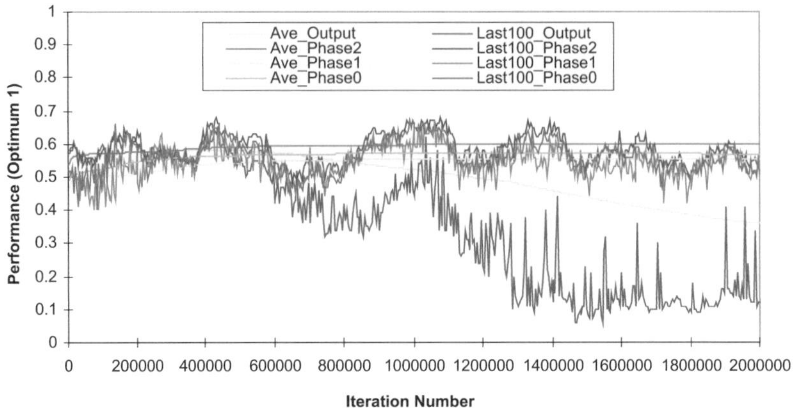

Fig. 5. Training performance of the industrial LCS with pinching real data

The output population of the pinching data set (see Fig. 5) stagnated resulting in the loss of performance from a million iterations onwards. This was due to the LCS finding good, but rare rules that were not deleted as more general rules of equivalent accuracy were not possible to discover. Thus the lack of exact matching rules in the output phase caused the lack of prediction, rather than incorrect prediction from matching rules. The LCS had experienced similar situations in the training set but even though it could reject unimportant parameters through generalisation it did not abstract the similarities in the important parameters. Any slightly novel situations, as often occurs in industry, was not matched, e.g. Rule '78 cm strip width, 76 cm guide setting' would not match a message '80 cm width, 79 cm guide.'

The real environment can be processed iteratively to give the LCS a greater chance of discovering good rules. The data collection can be directed at areas where faults were more frequent, such as thin gauges. This reduces the search space leading at least to improved efficiency. It can also lead to improved effectiveness, as instead of families of rules being formed one rule can be determined. For example, the same fault for low grades does not have to be separated into each grade.

4.2 Discovery Training Results

Decoding of the training for the pinching data set is shown truncated in Table 2. Including all 64 parameters in 100 rules would occupy 22 pages in this font, making decoding difficult. This task could be made easier by automatically removing conditions that are wild. Similarly, removing and reporting conditions that are the same for all output states would reduce investigation time.

Table 2. Results from pinching dataset

No.	Fertility		Parameters						Output	Spec.	Life	Child
1	1	Max	32,204	1,570	113.1	38.3	100.4	300	[0]	23	99,996	5
		Min	20,004	1,127.5	42.2	28.6	27.1	0				
		Resol-ution	1,562.5	81.2	9.8	1.2	12	30				
2	0.26		35,629	1,488.8	129.8	39.7	140.1	270	[0]	9	11,400	0
			20,004	853.8	48.8	27.9	32	0				
			1,562.5	81.2	9.8	1.2	12	30				
3	0.65		35,629	1,553.7	137	39.7	140.1	300	[0]	10	43,400	0
			20,004	839.2	50.2	28.6	20	0				
			1,562.5	81.2	9.8	1.2	12	30				
4	0.86		35,629	1,553.7	137	39.5	140.1	300	[0]	10	99,400	1
			20,004	839.2	50.2	29.8	32	0				
			1,562.5	81.2	9.8	1.2	12	30				
5	0.88		33,766.5	1,472.5	118.1	39.5	88.4	300	[0]	15	81,400	2
			20,004	839.2	48.8	28.6	32	30				
			1,562.5	81.2	9.8	1.2	12	30				
6	0.1		35,629	1,570	127.2	38.3	128.1	300	[1]	10	1,400	0
			20,004	839.2	40.4	28.6	20	30				
			1,562.5	81.2	9.8	1.2	12	30				
7	0.13		35,629	1,570	122.9	39.5	140.1	300	[1]	5	3,400	1
			20,004	758	59.7	27.9	20	0				
			1,562.5	81.2	9.8	1.2	12	30				
8	0.12		34,066.5	1,553.7	137	39.7	140.1	270	[1]	8	3,400	2
			20,004	758	50.2	27.9	32	0				
			1,562.5	81.2	9.8	1.2	12	30				
13	0.12		35,629	1,570	127.2	38.5	140.1	300	[1]	9	3,400	2
			20,004	839.2	60	27.9	20	30				
			1,562.5	81.2	9.8	1.2	12	30				
14	0.1		35,629	1,570	120	38.3	128.1	270	[1]	11	1,400	0
			21,566.5	853.8	39	27.9	20	0				
			1,562.5	81.2	9.8	1.2	12	30				
9	0.13		35,629	1,488.8	122.9	38.4	140.1	300	[2]	13	5,400	0
			20,004	935	60	29	20	30				
			1,562.5	81.2	9.8	1.2	12	30				
10	0.12		35,629	1,570	137	38.5	140.1	300	[2]	11	5,400	0
			20,004	839.2	39	29	20	0				
			1,562.5	81.2	9.8	1.2	12	30				
11	0.12		34,066.5	1,570	137	39.7	140.1	300	[2]	7	5,400	2
			20,004	758	39	27.9	20	30				
			1,562.5	81.2	9.8	1.2	12	30				
12	0.3		35,629	1,570	127.2	39.7	140.1	300	[2]	4	35,400	14
			20,004	839.2	50.2	27.9	20	30				
			1,562.5	81.2	9.8	1.2	12	30				
15	0.13		35,629	1,570	137	39.5	140.1	300	[2]	5	7,400	2
			20,004	758	39	27.9	20	0				
			1,562.5	81.2	9.8	1.2	12	30				

Discussion of the learned knowledge with 'experts' in steel rolling [9] showed that the industrial LCS did learn relevant information. Much of the information confirms existing Rules of Thumb, Heuristics and Empirical information. This is important as it builds confidence in the ability of the LCS. Examples of common knowledge are that exotic grades, wider widths and lighter gauges are more likely to cause problems such as pinching. This assists in further pre-processing the training set as easy to roll slabs can be separated from 'problem' slabs to determine more subtle mill performance.

The industrial LCS also discovered uncommon knowledge, in so much as the information was not widely available. A good example was that high Crown strip was unlikely to pinch. This was considered accurate as the crown has a centering effect when passed through rolls, similar to a pulley and belt effect. Although experienced operators knew this effect, the LCS could provide a means of discussion and harmonisation of plant knowledge throughout the workforce.

4.3 Advisory Performance

The prediction accuracy of the industrial LCS was only 60%, which was not sufficient for operators to be confident in the LCS. Although the knowledge contained in the data set was unlikely to completely describe the mill to enable 100% prediction, it was considered that enough was present for this testing method to achieve significantly better results. The data itself caused some of the problems for the predictive accuracy due to the redundancy of some parameters, the lack of volume and suspected noise. However, the data inadequacies were not thought to be the major problem with LCS performance.

The increased dimensionality of the domain overwhelmed the LCS as it could not identify useful general areas in the huge search space. The modality caused by including all relevant parameters, instead of separating training on specific product groups, restricted performance by requiring high specificity rules to be formed. The greater the specificity required in a rule, the more time on average it would take to form from disparate building blocks.

The scalability of the industrial LCS to this domain is encouraging, as limited training occurred. However, improved searching was needed to find better general rules; especially for the more obscure fault levels. The capture and retention of partly formed rules that form good building blocks was a major problem. These rules were less accurate than rules that dealt with specific instances, so were gradually removed from the populations. The LCS required more time and space to discover accurate general rules.

4.4 Summary of Industrial LCS

A visual summary of the results is shown in Fig. 4 where the performance of the industrial LCS reaches optimum compared with the non-training basic

LCS. The necessity for a balanced system with all methods complementing each other is shown in Fig. 3. Table 1 shows that the diagnosis is correct for known data and can identify hierarchy forms.

A typical performance of the industrial LCS with real data is shown in Fig. 5. Performance was not sufficient for plant confidence, possibly caused by poor data and/or inadequacies in the technique. Improvement in training performance occurred with useful information being discovered when the results are analysed.

Attempting to apply an LCS to industrial data in 1994–1998 was too ambitious, but did lead to some insights into both LCS and how they may be developed.

- Real alphabet [10, 11]: the majority of LCS used in data-mining do not use binary representations. Either a real alphabet (often single value with range) or data mapped onto specified ranges is used. For example, mapping temperature data having values 100–900 could map onto 10 ranges from 1 to 10, allowing the precision of each range to be tailored. The mapped range is then specified in bits in the ternary alphabet, e.g. 400 is in range 5 which is represented by 1111100000.
- Phases of training: single population LCS still continue to be the norm due to their simplicity. The GALE system [12] does implement multiple populations, which has shown benefits by autonomously determining correct LCS parameters set up. However, many populations are required thus increasing computational time and the connectivity between populations is not sequential.
- Evaluation limits are now common [13], but the child limit is rare. Subsumption deletion and Numerosity [14] are used instead and assist in removing similar information from the rule-base in XCS-type systems. However, as the number of attributes in a classifier increases, the likelihood of subsumption and numerosity occurring decreases, and the computational time required for these methods increases.
- Rule discovery: although Panmictic rule discovery has shown to be useful in allowing training of the whole rule-base regardless of action base rates, most modern LCS use niche based approaches as this gives an automatic generalisation pressure. Deleting rules becomes equally as important as discovering rules as this is the method used to balance base rates. Until large condition data sets (>32 parameters) are routinely addressed, then there will be no need for panmictic rule discovery or morphing with its inherent local optima problems.
- Split fitness measure: generally a single fitness measure is still used for selection for effecting and selection for mating as it is simplest to implement. More work is required to prove the benefits of a split fitness measure. ‘Accuracy of prediction’ is the preferred fitness measure, but ‘strength’ based approaches are being revisited [15, 16].

- Removal of taxation and genotypic specificity measures for mating: modern LCS also do not use taxation or genotypic specificity as they increase complexity with no benefit.

Applying LCS to industrial domains in 2007 is still not a trivial task, although the likelihood of beneficial results in prediction and diagnosis of plant has much improved. Brief comments are given below as rules of thumb:

- Reduce the complexity of the domain as much as possible without introducing bias or removing useful information, e.g. consider one type of product and one fault at a time.
- Collect as much good quality data as possible, whilst balancing base rates of actions to a practical amount.
- Consider an LCS where transparency and co-operating rules are required and high modality, parameter interaction and/or epistasis are present in the data.
- Choose the type of LCS; Pittsburgh for simple or known rule structures, Michigan for flexibility.
- Select appropriate LCS methods, using the anticipated accuracy of developed rules as a guide. High accuracy will favour XCS type systems.
- Real-alphabets are beneficial, with parameter range tailoring as necessary.
- Niche based rule discovery with mutation and crossover operators works well with moderate parameter numbers (roughly < 32) and balanced base rates of action.
- Set up the internal LCS parameters through iteration or using multiple populations.
- Test sets are useful to identify overfitting of rules, especially with low volumes of data.
- Pass through rules are useful when describing a function.

Splitting the population into three phases allows iLCS to discover general and specific building blocks of information in the search phase. Combining the small blocks of information into complete rules becomes increasingly difficult as the size of the domain increases, with the dedicated combine phase in iLCS facilitating this task. Whereas in a simple LCS, by the time general rules have been formed, the building blocks of information needed for specific rules have already been removed from the population. The different types of rules within the combine phase test alternative hierarchy structures, which is required if sub-optimum hierarchies, such as homomorphic or "flat", are not to form. Once good rules survive, proving their long-term use, they can be transferred to the stable population, where the information can be exploited through exact matching and best classifier selection.

The total population size is equivalent to a standard single population, but with analysis only necessary on the compact stable population. A slight increase in training time is noted due to evaluating three outputs, instead of one, for each environmental message, but this is acceptable within the overall training time.

Life limits are essential for introducing new (including transferred) rules into a stable population or hierarchy, as their appropriate level can be determined without interrupting the existing structures

The fertility fitness measure is dedicated to rule discovery, so can directly measure the benefit of mating a classifier. Alternative schemes utilise the same measure (e.g. strength, accuracy or prediction of payoff) in selection for effecting as well as selection for mating. When developing complex hierarchies, which are likely to include initially underdeveloped branches, a single fitness measure may ignore appropriate, but weak, classifiers. This is acceptable in exploiting information, but not in exploring for improved overall performance. Fertility functions by estimating a classifier's usefulness compared with its neighbouring classifiers. Inaccurate classifiers have low fertility, unless there are no alternatives. Similarly, accurate classifiers have low fertility if they occur in less match sets than similar classifiers of equal accuracy. The preference measure could use any standard fitness measure. In this domain the accuracy of the classifier is used as payoff is the same for all actions, so using predicted payoff provides no additional information.

Real-alphabets are essential for transparency in industry, but they also improve boundary definition in continuous domains. Fine tuning rule boundaries can occur through mutation and resolution tuning, but is often achieved through the result of appropriate crossovers, where upper and lower attribute bounds are swapped. Adjusting the upper/lower bounds allows a single rule to describe a range, whereas ternary-alphabet may require more than one rule to complete the range.

5 Abstraction

The process of abstraction can be likened to Information Processing Theory [17] (a branch of Learning Theory), which suggests that humans have the ability to recognize patterns in data and chunk these patterns into meaningful units. The individual patterns do not necessarily remain in a memory store due to the holistic nature of the individual patterns. However, the chunks of meaningful information remain, and become a basic element of all subsequent analyses.

The Abstraction algorithm needs to perform this "chunking", for the individual patterns created by a learning system, see Fig. 6 [18]. The learning system selected was the XCS implementation of the Learning Classifier System concept as it has been shown to produce accurate and maximally general rule sets [14]. The LCS concept was derived from work by Holland [1] on developing artificial central nervous [cognitive] systems. Much past work has focused on improving learning performance, but recent work has revisited its cognitive abilities [19]. LCS use evolutionary computation to produce maximally general compact production rules, thus are a suitable technique to form base rules.

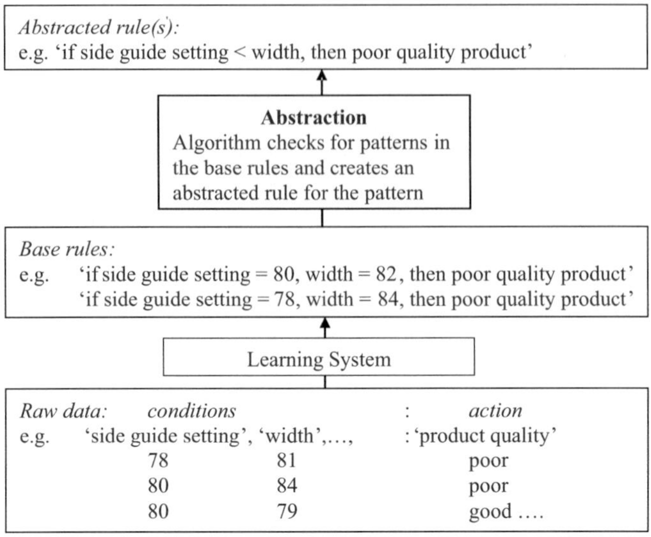

Fig. 6. Abstraction from data to higher order rules

The first step in developing the Abstraction algorithm is to select a suitable test domain. Games, such as Chess or Connect 4, offer a useful environment in which to test an algorithm's ability to find patterns within the data set and, more importantly, perform abstractions. Games are useful as they offer an environment that is well studied (meaning performance is easy to gauge), competitive, turn based (meaning time is not an important factor) and finite (although very large in most cases).

The goal of this project was to create an Abstraction algorithm that would generate rules for and play the game of Connect 4, due to this domain's scale, multi-step and non-deterministic properties.

5.1 Abstraction Algorithm Overview

The Abstraction algorithm was designed to work upon the rules generated by the LCS. Abstraction is independent of the data itself. Other methods, such as the standard coverage operator [7], depend directly on the data. Crossover and mutation depend indirectly on the data as they require the fitness of the hypothesized rules, which is dependent on the data. Abstraction is a higher order method, as once good rules have been discovered; it could function without the raw data being available.

The abstraction attempts to find patterns in the rules that performed best within the LCS. Having found a pattern common to two or more of the LCS rules, the Abstraction algorithm is to generate a new rule in the abstracted population based solely on this pattern. This allows the pattern to be matched when it occurs in any state, not just the specific rules that exist within the LCS.

Not all of the rules generated by the LCS are worthwhile and therefore the Abstraction algorithm should not be run upon all of the rules within the LCS. The domain is noiseless, so the parameters chosen to govern the testing of rules for abstraction were the conditions that a rule must have a 100% win score and a 100% accuracy. Therefore the rules abstracted by the Abstraction algorithms should only be rules that lead to winning situations.

The main mechanism that allowed the abstraction to perform was a windowing function that was used in rule generation as well as rule selection (when it came to choosing an abstracted rule to play). The windowing function acted as a filter that was passed over the 'good' rules generated by the LCS. This filter would compare two rules at a time for similarities that could lead to abstracted rules.

The windowing function worked in all directions on the board, horizontally, vertically and in both diagonal directions. The window size was set to 4 space/counters (eight characters in terms of the board representation). However code allowed for a window size of between 4 and 6 spaces/counter (8–12 characters in terms of the board representation), any greater than a window size of 6 and the vertical and diagonal windows no longer fit on the board.

Any match that is found is turned into an abstracted rule, each rule had eight characters (assuming a window size of 4) to represent the pattern occurring on the board. Each rule also had to be assigned a move to play whenever that rule was used. The move assigned was always chosen from one of the playable spaces within the pattern. An example rule is '10,10,10,00:11', which translate to 'if three red counters in a row and payable space in the next position, then play in the next position'. All rules entered the abstracted population with a win and accuracy of 50.

Several limitations were placed upon what was considered a valid match for the Abstraction algorithm, including ignoring all unplayable areas. A valid pattern had to contain at least one playable space and no more than two playable spaces. Patterns without a playable space are useless because rules as they offer nowhere for a move to be played. The second limitation placed upon the abstraction process was that a valid rule could have a maximum of one unplayable space. This helps limit the generation of "empty" rules. Figure 7 shows an example of two windowing functions finding a match and generating an abstracted rule.

As with the LCS, the Abstraction algorithm also had a GA that was run upon the population to generate new rules. It had a single point crossover and mutation; however it had no deletion algorithm as all the abstraction rules were kept. Duplication was prevented through a duplication check that was made each time a rule was to be inserted into the rule-base, including those generated by crossover and mutation.

A LCS can function alone, but the Abstraction algorithm cannot function without a rule-base to work on; hence it needs an LCS to function alongside it. How the two are combined and work together is detailed in this section.

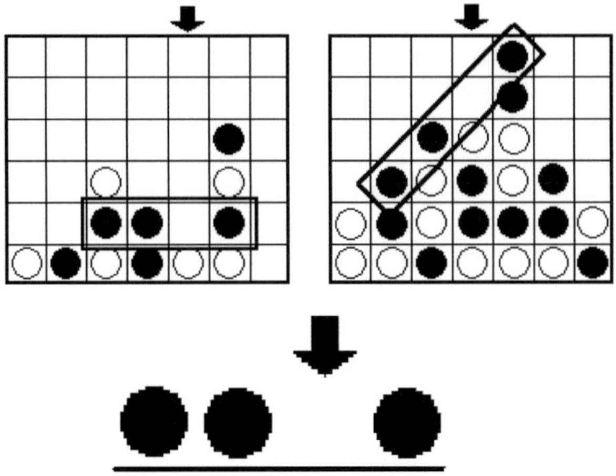

Fig. 7. Example of Abstraction algorithms generating a new rule

When the LCS with abstraction needs to play a move, the system searches the board for any matches within its abstracted rule set. The board is searched by passing the windowing function over the board (horizontally vertically and diagonally). A rule is then chosen out of all matched rules. When exploiting the rule with the best win score is chosen, whilst when exploring a roulette wheel based upon accuracy is used.

The chosen abstracted rule also has a move associated with it, however unlike the LCS rules the move does not relate directly to the board. With a window size of 4 counters the rule could occur anywhere on the board, horizontally, vertically or diagonally. Therefore an extra calculation is required to translate the abstracted rules' move into the corresponding move on the actual board.

If no abstracted rule is found after the initial search of the board state, then control of playing the move is handed to the base LCS.

6 Results

The following section details the results found during the trials of the LCS and Abstraction algorithm. Initial trials investigated the difficulty of the problem domain with standard Q-learning and XCS techniques. Preliminary tests of the Abstraction algorithm with XCS were followed by tests of the Abstraction algorithm with a modified XCS (mXCS), see Sect. 6.2, where the reinforcement learning complemented the abstraction. The use of abstraction as the training progressed was investigated. During these tests, each system was trained for 20,000 games against an opponent that played randomly. Finally, the robustness of the Abstraction algorithm to changes in the domain was tested by increasing the difficulty of the opponent.

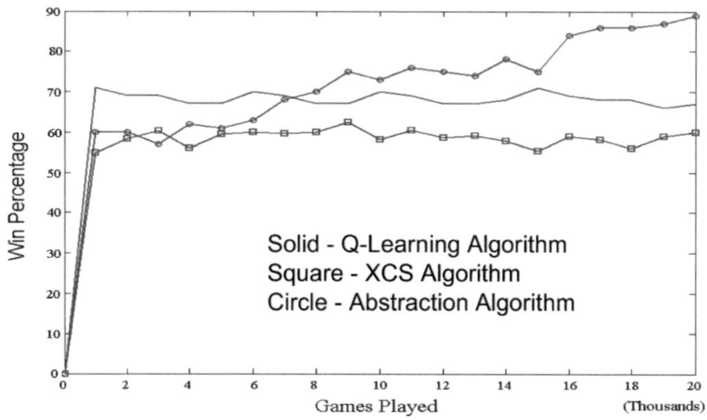

Fig. 8. Graphs of win percentages for three learning algorithms

6.1 Q-Learning and Standard XCS

The Q-Learning Algorithm performed well in the initial 20,000 games (see Fig. 8), achieving an average win percentage of 69%. However, there was no progress in the wins as the 20,000 games progressed, with the win percentage always remaining at around 69%. This exhaustive search nature of the algorithm meant it took several weeks of computation on a 3 GHz PC.

The XCS performance trend was similar, with an average win percentage of 62% reached quickly, but no further improvements. Analysis of the rules showed that they had become trapped in local optima. A few specific strategies had been learnt, such as initially trying to build a column of counters in a given column. However, if this column happened to be blocked, then the overall strategy failed.

When the Abstraction algorithm was included in the standard XCS the performance did not improve. This was because the XCS did not find sufficiently accurate rules for the Abstraction algorithm to be triggered.

The random nature of the opponent meant that a sometimes good strategy, such as build a column of counters in row one, was occasionally blocked. The prediction is updated by the Widrow-Hoff delta rule, which severely penalizes an incorrect prediction. Thus the accuracy of prediction never reaches a high stable level. Adjusting the learning rate (β) within the range 0.1–0.6 did not improve the performance. Reducing the threshold of what the Abstract algorithm considered as accurate to 85% accuracy of prediction also did not improve performance.

6.2 Modified XCS With and Without Abstraction

The Widrow-Hoff delta rule was replaced by a simple reinforcement learning update (+2 if correct, −2 if incorrect, limited to the range 0–100), where recency had a much reduced effect. This enabled mXCS to produce rules considered accurate enough to be abstracted. Although the rules were similar to those produced by XCS, the learning was more gradual, which prevented good rules from being replaced due to low accuracy of prediction.

When the Abstraction algorithm is added, a significant improvement is noted after 8,000 trials as the performance increases to 90%. This compares favorably with both Q-learning (69%) and standard XCS (62%), see Fig. 8.

During testing the rules that the Abstraction algorithm produced were observed and an interesting pattern arose in the order in which the abstractions were discovered. In early generations no abstracted rules are found, whilst mXCS attempts to establish a set of good rules that have a win and accuracy of 100. The first abstracted rules found are not rules for a direct win (i.e. three in a row and play in the fourth). The first rules that emerge are those rules that cause a three in a row situation with an empty playable fourth space.

Learning to form three in a row followed by learning to form four in a row is a novel example of incremental learning. Intuitively, it could be expected that learning to form four in a row, which is closer to obtaining the reward, would be achieved first. Incremental learning is hypothesized to be an important cognitive ability [8].

Whilst there is no direct feedback from the abstraction rule-base to the mXCS rule-base, it is possible to see them evolve together and there is a definite dependency between the two. With the introduction of abstracted rules to make three in a row, this is likely to occur far more often (as abstracted rules take preference over mXCS rules). With three in a row occurring more often, mXCS has more opportunities to conceive of rules that directly give a win. Therefore with more winning rules the Abstraction algorithm is more likely to discover further abstracted rules that lead to a direct win, greatly boosting the winning ability of the algorithm.

6.3 Effect of Abstraction

The use of abstracted rules as training progresses can be monitored, see Fig. 9. The combined system always plays a matching abstracted rule in preference to a matching base rule. After 8,000 trials the base rules were accurate enough to allow abstraction to start. Once abstraction had started, the performance of the system continued to improve beyond that of standard XCS and Q-learning (see Fig. 8). A further 8,000 trials occur where the system uses a combination of both base and abstracted rules. After this period the system just uses abstracted rules in its decision-making. Small improvements in performance occurred due to the action of the genetic algorithm in the abstracted population.

The random opponent still defeats the system in 10% of the games when it chances upon a good strategy. As there are multiple positions for good strategy is to occur in, the system is rarely presented with them, which makes

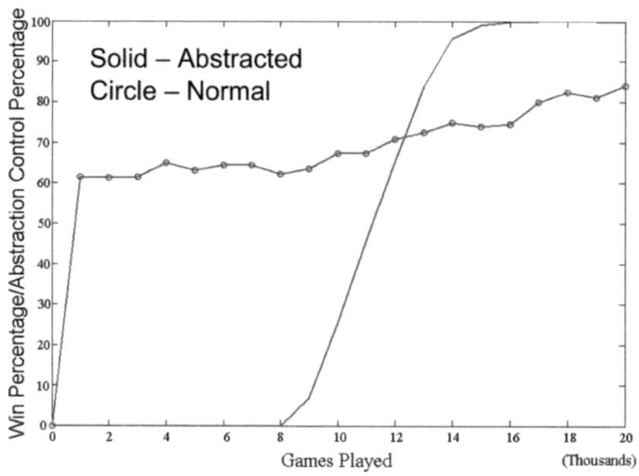

Fig. 9. Graph of percentage base rules versus abstracted rules as training progresses

them difficult to learn. In order to determine the robustness and scalability of the techniques the difficulty of the opponent was increased.

6.4 Robustness of the Systems

The opponent could now block a potentially winning three in a row state. The system has to learn to create multiple win situations. This is a significantly harder problem, especially as the opponent could win either randomly or in the act of blocking, which halts the game. All algorithms perform poorly as all win percentages were under 20% and no accurate rules are present. If no good base rules are found, then the Abstraction algorithm will not start.

Instead of training from the start with the harder opponent, it was decided to train first with the simple opponent and then switch to the harder opponent, see Fig. 10. After the switch, standard XCS performed better than the Q-Learning Algorithm, achieving a win percentage of 15%, it should be noted that the performance was less than the Q-Learning algorithm during the first 20,000 games. Analysis of the Q-Learning algorithm testing showed that progressive training, from the easiest to the hardest opponent, caused it to get stuck in a local optimum with a win percentage of only 11%. The generality and adaptability of the standard XCS algorithm enables it to switch opponent without penalty.

The performance of the Abstraction algorithm was significant. Not only did it outperform standard XCS and Q-learning (53%, compared with 15% and 11% respectively), but it performed significantly better then when it had been trained only on the harder opponent (53% compared with 19%). This is a good example of incremental learning, where it is necessary to build up the complexity of the problem domain.

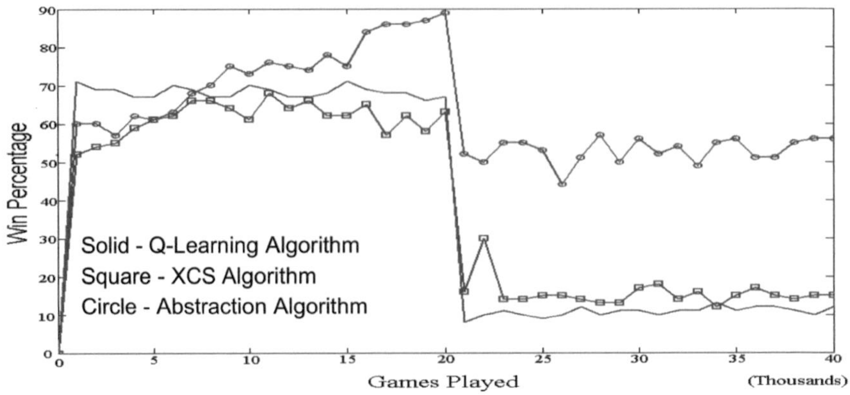

Fig. 10. Change in opponent at 20×10^3 games played

7 Conclusions

An industry focused Learning Classifier System, iLCS, has been developed for application to complex domains that place a severe demand on exploration, exploitation and balancing these often conflicting needs. The iLCS has improved on simple LCS through the use of training phases, life limits, tailored rule discovery, separate fitness measures for mating (fertility) and effecting (preference), removal of taxation and use of a real-alphabet. Importantly, iLCS offers a different approach to developing LCS than the advanced XCS-type systems, which may prove necessary as more complex domains are considered.

Utilising a balance of existing and novel methods, iLCS achieved the desired performance level on simulated data, with small niches being discovered and maintained. Information is learnt from complex real-data, which can be used to direct an engineer's search for causes of problems. However, more work is required to improve prediction accuracy for domains characterised with high dimensionality, epistasis, multimodality and lack of separation where higher-level rules are required.

An Abstraction algorithm has been demonstrated to produce appropriate higher-level rules in a complex toy problem, where learning is now also based on discovered knowledge rather than just the environmental data. LCS are a complex evolutionary technique, where selection of appropriate strategies, such as abstraction, for a given domain will lead to benefits for industry.

References

1. Holland, J. H.: Adaptation in Natural and Artificial Systems. University of Michigan Press, Ann Arbor, MI (1975 and 1992)
2. Wilson, S. W.: ZCS: a zeroth level classifier system. Evolutionary Computation (1994) 2(1):1–18

3. Wilson, S. W.: Classifier fitness based on accuracy. Evolutionary Computation (1995) 3(2):149–175
4. Booker, L. B., Goldberg, D. E., Holland, J. H.: Classifier systems and genetic algorithms. Artificial Intelligence (1989) 40:235–282
5. Wilson, S. W.: Structure and Function of the XCS Classifier System. Lecture Presented at the Rowland Institute for Science, Cambridge, MA. Available at http://netq.rowland.org (1998)
6. Goldberg, D. E., Horn, J., Deb, K.: What Makes a Problem Hard for a Classifier System? Available from ENCORE in file CFS/papers/lcs92-2.ps.gz (1992)
7. Bull, L.: Applications of Learning Classifier Systems. Springer, Berlin Heidelberg New York (2004)
8. Venturini, G.: Adaptation in dynamic environments through a minimal probability of exploration. In: Cliff, D., Husbands, P., Meyer, J. A., Wilson, S. W. (eds) *From Animals to Animats 3: Proceedings of the Third International Conference on Simulation of Adaptive Behaviour*, MIT Press, Cambridge, MA (1994) pp. 371–379
9. Jones, A.: A7, Information Systems, British Steel Strip Products (1998) (personal communication)
10. Browne, W. N. L.: The Development of an Industrial Learning Classifier System for Application to a Steel Hot Strip Mill. Doctoral Thesis, University of Wales, Cardiff (1999)
11. Wilson, S. W.: Get real! XCS with continuous-valued Inputs. In: Lanzi, P. L., Stolzmann, W., Wilson, S. W. (eds) *Learning Classifier Systems, From Foundations to Applications*. Lecture Notes in Computer Science, Vol. 1813, Springer, Berlin Heidelberg New York (2000) pp. 209–222
12. Bernadó, E., Llorà, X., Garrell, J. M.: XCS and GALE: a comparative study of two learning classifier systems on data mining. In: Lanzi, P. L., Stolzmann, W., Wilson, S. W. (eds) *Advances in Learning Classifier Systems*. Lecture Notes in Artificial Intelligence, Vol. 2321, Springer, Berlin Heidelberg New York (2001) pp. 115–133
13. Butz, M., Wilson, S. W.: An algorithmic description of XCS. Soft Computing: A Fusion of Foundations, Methodologies and Applications (2002) 6:162–170
14. Lanzi, P. L.: A study of the generalization capabilities of XCS. In: *Proceedings of the Seventh International Conference on Genetic Algorithms*, Morgan Kaufmann, USA, (1997) pp. 418–425
15. Wada, A., Takadama, K., Shimohara, K.: Dual-structured classifier system mediating XCS and gadient descent based update. *An International Workshop on Learning Classifier Systems 2006*, proceedings to appear
16. Kovacs, T.: Strength or accuracy? Fitness calculation in learning classifier systems. In: P. L. Lanzi, W. Stolzmann, S. W. Wilson (eds) *Learning Classifier Systems: An Introduction to Contemporary Research*. Springer Berlin Heidelberg New York (2000) pp. 143–160
17. Miller, G. A.: The magical number seven, plus or minus two: some limits on our capacity for processing information. Psychological Review (1956) 63:81–97
18. Browne, W., Scott, D.: An abstraction algorithm for genetics-based reinforcement learning, GECCO 2005. In: Hans-Georg Beyer, Una-May O'Reilly, Dirk V. Arnold, Wolfgang Banzhaf, Christian Blum, Eric W. Bonabeau, Erick Cantu-Paz, Dipankar Dasgupta, Kalyanmoy Deb, James A. Foster, Edwin D. de Jong, Hod Lipson, Xavier Llora, Spiros Mancoridis, Martin Pelikan, Guenther R.

Raidl, Terence Soule, Andy M. Tyrrell, Jean-Paul Watson, Eckart Zitzler (eds) *Proceedings of the 2005 Conference on Genetic and Evolutionary Computation*, Washington, DC, USA, (2005) pp. 1875–1882
19. Butz, M.: Rule-Base Evolutionary Online Learning Systems: Learning Bounds, Classification and Prediction. PhD Thesis, University of Illinois, Illinois (2004)

Distributed Learning Classifier Systems

Hai H. Dam[1], Pornthep Rojanavasu[1,2], and Hussein A. Abbass[1], and Chris Lokan[1]

[1] Artificial Life and Adaptive Robotics Laboratory, School of Information Technology and Electrical Engineering, The University of New South Wales, Canberra, NSW, Australia, z3140959@itee.adfa.edu.au, s3205770@itee.adfa.edu.au, abbass@itee.adfa.edu.au, cjl@itee.adfa.edu.au, http://www.itee.adfa.edu.au/~alar

[2] Department of Computer Engineering, Faculty of Engineering, Research Center for Communication and Information Technology, King Mongkut's Institute of Technology Ladkrabang, Thailand

Summary. Genetics-based machine learning methods – also called learning classifier systems (LCSs) – are evolutionary computation based data mining techniques. The advantages of these techniques are: to provide rule-based models that represent human-readable patterns; to learn incrementally, capable of adapting quickly to any changes in dynamic environments; and some of them have linear 0(n) learning complexity in the size of data set. However, not too much effort has yet been put into investigating LCSs in distributed environments. This chapter will scrutinize several issues of LCSs in distributed environments such as knowledge passing in the system, knowledge combination methods at the central location, and the effect on the system's learning accuracy of having different numbers of distributed sites.

1 Introduction

Pervasive computing has opened a new era in technology and communications where people can access and/or transfer data around the world in real time. Instead of tons of books, a huge amount of information can be preserved within small and affordable electronic systems. Along with fast and convenient ways to access the outside world, huge amounts of electronic data are also generated with a greater frequency than at any time in the past. Data mining, the process of discovering novel and potentially useful patterns in the data [11], has become an effective method to discover the tacit knowledge hidden in such overwhelming databases.

Nowadays, most databases in large organizations are distributed physically in many locations due to the trend of globalization. For example, a company might have multiple branches placed in many cities, states, countries, etc. From the management's perspective, the data generated from multiple locations need to be integrated into a single coherent knowledge base for the future

H.H. Dam et al.: *Distributed Learning Classifier Systems*, Studies in Computational Intelligence (SCI) **125**, 69–91 (2008)
www.springerlink.com

decision making. However, with the large amounts of data generated daily at each location, it is not possible to transfer all the data to a central location for normal data mining due to security issues, limited network bandwidth, and even because of the internal policies for some organizations. Distributed Data Mining (DDM), an extension of data mining techniques in distributed environments, was introduced to tackle this problem. The primary purpose of DDM is to discover and combine useful knowledge from databases that come from multiple sources [21]. Even if the data is not physically distributed, DDM can be used as a parallel system to speed up the mining process.

In many data mining problems, the ability to understand the knowledge discovered by the mining algorithms is sometimes as important as obtaining an accurate model. For instance, a company might want to profile customers' expenditures in terms of their consumption, services, location, income, season, etc. The relationship between those features with regards to customers' spending habits can answer questions regarding their purchase behaviors. This sort of information would be valuable for managers to identify the best segments to target their marketing campaigns.

Discovered patterns from mining algorithms can be represented in different forms such as neural networks, decision trees, or a set of rules. It is potentially easier to understand patterns represented using the latter form than those for example represented using the neural network. In fact, it is possible to obtain a set of rules from the trained neural network but the process requires a special rule extraction algorithm. This post-process generally needs to run off-line and therefore cannot provide the knowledge on the fly as required in many stream data mining applications.

Genetics-based machine learning systems [13] are evolutionary rule-based systems, the most common of which is known as Learning Classifier Systems (LCSs). In LCSs, a set of rules (called the population of classifiers) is evolved to guide the system to achieve some tasks in an arbitrary environment. LCSs employ two biological metaphors: *evolution* and *learning*. The evolutionary component plays a key role in discovering novel and potentially useful rules, while the learning component is responsible for assigning credit to rules in the population based on their (estimated) individual contribution to achieving the task. Thereby, learning guides the evolutionary component to move towards a better set of rules.

Work on LCSs normally falls in one of two categories: the Pittsburgh [22] and the Michigan [17] approaches. The major difference between the two approaches is that an individual of the Pittsburgh approach is a set of rules representing a complete solution to the learning problem. In contrast, an individual of the Michigan approach is a single rule that represents a partial solution to the overall learning task. Thus, the Michigan and the Pittsburgh systems are quite different approaches to learning. Many studies have compared the performance of the two approaches on several data mining problems in order to determine circumstances in which one approach would perform better than the other [1, 2, 20]. Their results revealed in some cases that the Pittsburgh approach is more robust, but it is also computationally very expensive compared to the Michigan method.

Our current research focuses on mining data streams, where data instances are continuously arriving in real time. Due to memory constraints when faced with an infinite stream of data, each instance is presented to the system at most once before being discarded. We consider that Michigan-style LCSs are more suitable in this domain, due to their abilities to learn incrementally on the fly and their smaller space and time complexity.

Many models of Michigan-style LCSs have been introduced in the last two decades. Recently, the accuracy-based models have captured most attentions from researchers in the field because of their good performance. XCS [23, 24] is the first system of this type, which works for both supervised learning and reinforcement learning problems. UCS [4] is a derivation of XCS that specializes on classification tasks. Since we are working at classification (supervised learning) problems, UCS was chosen as our baseline LCS learner.

In 2005 we proposed a framework using XCS in distributed computing environments, called DXCS [8, 9]. In this framework, the distributed and central sites are described in terms of clients and a server. To validate the system, we compared it with a centralized XCS, which basically transfers all data from remote sites to a central location for centralized data mining. The results revealed that DXCS is competitive as a distributed data mining system for two reasons. First, DXCS has similar accuracy to the centralized system. Second, the amount of data needed to be transferred to the central location is reduced enormously in DXCS compared to a centralized XCS system.

In this chapter, the framework of DXCS is extended for other LCSs and called DLCS. We investigate DLCS on several issues related to knowledge passing in the system, the amount of data transmission, etc. This chapter aims to answer the following research questions.

- Do different methods of knowledge combination at the server affect the learning? In this chapter, the knowledge probing and majority voting approaches are employed to combine the decisions from local models.
- Does the number of clients in DLCS affect the convergence of learning at the server? The previous papers only experimented with DXCS with two clients. In theory, adding more clients provides more knowledge for the server. We hypothesize that the server is able to learn faster when adding more clients.
- How to transfer the knowledge in DLCS in order to speed up the convergence? In theory, clients need to update their knowledge at the server as frequently as possible. In our previous paper, we transferred the whole population to the server at a constant rate. However, we believe that sending only partial models would reduce the traffic, but can still maintain the predictive accuracy. Moreover, exchanging the knowledge between the clients (instead of limiting knowledge passing between clients and the server) would help to speed up the learning at clients and therefore at the server.

The chapter is structured as follows. The next section provides a short review of distributed data mining. Section 3 introduces UCS. Section 4 describes our proposed framework of DLCS. The experimental setup is explained in Sect. 5 followed by several discussion sections. Finally, the conclusion is presented in Sect. 9.

2 Distributed Data Mining

2.1 Overview

DDM is a relatively new area but it has been receiving much attention, especially in distributed computing environments where trust between sites is not always complete or mutual [18]. In many applications, data are privacy-sensitive, so that centralizing the data is usually not acceptable [12]. Therefore, DDM technology is adopted in many applications to reduce the transmission of raw data in the network and thus provide better protection to data.

According to Kargupta [19], DDM consists of two levels: the local level - extracting local knowledge at distributed sites; and the global level - combining, fusing, and resolving the local knowledge in order to build the global knowledge. The typical architecture of a DDM system is illustrated in Fig. 1. It normally contains three main phases: mining at the local sites; transmission of data among remote sites; and combining the local knowledge into a single coherent knowledge base. For example, each local branch in a company receives local data which would result in different learnt models. The central office of the company is responsible to combine these local models in order to gain an overall view of the company as a whole.

Distributed databases can be divided into two main categories: homogeneous and heterogeneous. In the former case the databases located at different sites have the same attributes in the same format (same data dictionary). In the latter case, each local site may collect different data, thus they may have

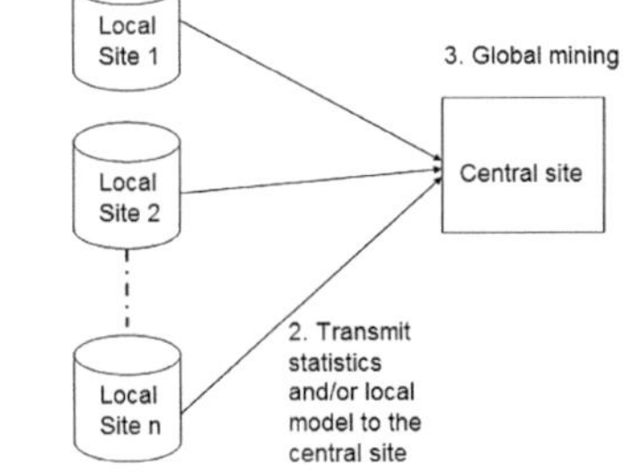

Fig. 1. A typical architecture of DDM

different numbers of attributes, types of attributes, formats of attributes, etc. The focus of this paper is on homogeneous databases.

2.2 Learning in Distributed Environments

There are three main schools existing in the literature for mining distributed databases. These are [7]:

- Combining data: Combine the distributed databases logically as a single large database during the learning process and allow communication of the intermediate results. The learned concept represents the distributed data and can be applied to incoming data instances.
- Combining local models: Learn at individual local sites independently, and combine the locally learned models to form a global concept. For example, the locally learned model can be some kinds of rules or decision trees. The global concept is the final rule set or decision tree made by the local models.
- Combining the predictive models: When receiving an incoming new instance, local models are used to predict the class then the locally learned concept in each local database is used to form a consensus classification. This strategy is to apply a classification technique on the local database to form some meta-ndata for further classification training. Thus we have multiple levels of classifiers in a distributed environment.

The first approach, or so-called centralized data mining, is the easiest to implement. It is either ineffective or infeasible, thought, for several reasons:

- Storage Cost: It is obvious that the central storage system needs to be very large in order to host all data from the distributed sites. Consider a supermarket with several stores. Data from local stores is sent to the centralized site daily, weekly, or monthly. A record at each store represents a single transaction of a customer, which contains information of payment details and products purchased. A small store might have only a few hundred transactions per day, but a large store might have an order of magnitude or more transactions per day. The size of data of a large store such as Wal–nMart is reaching the scale of gigabytes and terabytes daily and continuing to increase. Thus, central storage requires a huge data warehouse which costs enormously.
- Communication Cost: To transfer huge volumes of data to the central site might take an extremely long time and require huge network bandwidth. Since data at local sites is not static, their knowledge needs to be transmitted frequently in order to maintain up-to-date knowledge at the central site. An efficient algorithm needs to reduce the bottleneck of the transmission but at the same time has to update data as frequently as possible.

- Computational Cost: The computational time at the central site is much bigger than the aggregate cost of learning different models locally in parallel at local sites.
- Private and Sensitive Data: Sending raw data to the central site is not desirable since it puts privacy at risk, at either the central location or during transmission in the network. In many cases, for example, local branches might need to compete with each other; thus they may wish to limit data exchange to just the knowledge level related to a particular data mining exercise without sending all of their raw data.

The other two approaches (combining local models, and combining predictive models) seem more attractive to researchers for two main reasons. Firstly, the local model is normally much smaller than the raw data; sending the learnt model instead of the raw data thus reduces the traffic load in the network as well as the network bandwidth requirement. Secondly, sharing only the model, instead of the raw data, would give reasonable security since it overcomes partially the issues of privacy and security of the raw data.

Several algorithms have been introduced in the literature for combining local models. Hall et al. [15, 16] present a method to convert decision trees from distributed sites into a set of rules, and for combining those rule sets into a single rule set. Cho et al. [7] propose an alternative method that learns a single rule for each class at each local site. The rule of each class is shortened by an order of importance with respect to confidence, support, and deviation. The final rule set contains a set of the first n rules in the list. The most important part of this approach is to resolve conflicts while taking the union of the distributed rule sets.

A large fraction of DDM algorithms focuses on combining predictive models. This approach has emerged from empirical experimentation due to a requirement for higher prediction accuracy. Recently, several researchers treat distributed learning systems as a centralized ensemble-based method [10]. Several learning algorithms are applied at each local site, using separate training data to mine local knowledge. A new data point is then classified/predicted from the predictions of all local sites using ensemble methods such as stacking, boosting, majority voting, simple average, or winner-take-all methods. In general, DDM approaches apply ensemble methods to minimize the communication costs and to increase the system prediction.

Meta-learning methodology was introduced by Chan and Stolfo [6] in 1993, based on the idea of Stacked Generalization [26]. In this approach, each local site may employ a different inductive learning algorithm for constructing its local model. A meta-classifier is trained using data generated by the local models. This process is applied recursively to produce an arbiter tree, which is a hierarchy of meta-classifiers. This approach has been showed to achieve better performance in comparison to majority voting for fraud detection in the banking domain. An alternative method built on the idea of meta-learning is knowledge probing (KP), developed by Guo and Sutiwaraphun [14]. This method uses an independent dataset and does not build an arbiter tree. The output of a meta-learning system on this independent dataset, together with the label vector of the same dataset, are used to form a new training set for the learning algorithm that results in the final model. In our previous paper, we applied KP and achieved good performance. This paper will compare KP with a simple but effective approach, majority voting.

3 Learning Classifier Systems

LCS is a rule-based evolutionary learning classifier system, in which each classifier implements a partial task to the target problem. A typical goal of LCS is to evolve a population of classifiers $[P]$ to represent a complete solution to the target problem. LCS can be used either for reinforcement learning (e.g. XCS) or supervised learning (e.g. UCS) for evaluating classifiers in the population. This paper focusses on UCS (supervised classifier system).

Each classifier consists of a rule and a set of parameters. Each rule is made up of a *Condition* (the body of the rule), and an *Action* (the prediction of the classifier). The *Condition* embodies several environmental states, which the classifier may match to. The *Action* is a proposed outcome of the classifier if it is fired/activated. The crucial parameter associated with each classifier is its fitness F, which measures its goodness relative to the rest of the classifiers in the population. Another two important parameters are *numerosity* and *experience*. Each classifier in UCS is considered as a macro-classifier; that is, its rule is unique in the population. The *numerosity* parameter records the number of times that the classifier has been introduced in $[P]$ (mainly by a Genetic Algorithm (GA) operator). When a new classifier is introduced, the whole population is scanned through to check for its existence. If it does not exist, the classifier is added to the population; otherwise the numerosity parameter of the existing classifier is incremented by one. The *experience* parameter indicates how often the classifier has appeared in the match set; over time, this reflects its generality.

During both the exploration (training) and exploitation (testing) cycles, UCS repeatedly receives input from the environment. A match set $[M]$ is formed for each input, containing all classifiers in the population $[P]$ whose *condition* matches the input. Classifiers in $[M]$ will work together to decide on the system's outcome.

In the exploitation phase, a *prediction array* $[PA]$ is formed estimating the probability of each possible outcome in $[M]$ with regards to their fitness. The action with the highest probability in $[PA]$ is selected as the system's prediction.

The exploration phase is more complicated, as it involves both learning and searching. Since UCS is a supervised learner, a desired class/outcome accompanies the input. A *correct set* $[C]$ is formed, containing those classifiers in $[M]$ that have the same *action* as the input. If $[C]$ is empty, *covering* is applied, where a classifier that matches the input is created and assigned the same outcome as the input. UCS is an incremental learner, where knowledge is updated as more data becomes available. All parameters of the classifiers in $[M]$ are revised for each training instance, reflecting the system's response to new knowledge.

GA is invoked in $[C]$ if the average time since the last GA activation of classifiers in $[C]$ surpasses a user-defined threshold. Two parents are selected from $[C]$ with probability proportional to their fitness. Two offspring are generated by reproducing, crossing-over, and mutating the parents with certain probabilities. Offspring are inserted in $[P]$ if they are not subsumed by the parents. If the population size reaches a predefined limit, some classifiers are removed by voting within the population.

4 Distributed Learning Classifier Systems (DLCS)

This section describes the architecture of our proposed system, called DLCS. Similar to DXCS, DLCS is constructed of a number of clients and a server. Each client is placed at a local distributed site and is responsible for gathering the local knowledge. The local knowledge is organized to be transferred to the server, within a window frame. The server is responsible for preserving and combining these local models in order to form a descriptive model of the global environment.

4.1 The Framework of DLCS

Clients

Each client employs a complete LCS which is trained independently on its local database. A local model in DLCS is a population of classifiers learned by the local LCS. The population of classifiers at each client starts from an empty set and keeps evolving during the training process.

The data transmitted to the server consists of the local learned model (the population of classifiers) along with a set of data instances. The set of data instances is mainly used for training at the server if required.

Server

The server stores local models in the memory and avoids any modification of their contents. Local models are synchronized with the updated models newly arrived from their local sites. It allows the server to keep the most up-to-date knowledge of the global databases. Any change at the clients will be realized by the server through the updated model. The frequency of transmission affects the knowledge of the server. The more often local models are updated, the more up-to-date the server is.

The clients' models are aggregated at the server. Each input instance at the server is processed by all local models, which might result in different predictions due to different bias learned by clients. The final output is chosen between those predictions, using some knowledge fusion method. The next subsection will describe two existing methods for combining these predictions, which are used to test DLCS in this chapter.

4.2 Knowledge Combination at the Server

After receiving the updated models and training data from all clients, the server will combine the local models into a single coherent knowledge base. In our previous paper we applied knowledge probing technique in the distributed framework, and obtained promising results. In this chapter, we will compare knowledge probing against majority voting for combining local predictions at the server. We chose majority voting because it is an effective and very simple approach, which is widely used in the literature of ensemble learning.

Knowledge Probing

A simple LCS is employed at the server to learn the outputs from the local models. Clients also provide some data instances for training at the server. An algorithm to create a new training instance for the LCS is described as follows.

- Form New Inputs:
  - A set T of n training instances obtained from clients $T = \{(t_1, c_1); (t_2, c_2); ..; (t_n, c_n)\}$ where t is a training instance and c is an associated class.
  - A set L of m local LCS models $L = \{l_1, l_2, ..., l_m\}$.
  - A learning algorithm LCS on top of local models, which provides a final descriptive output of the server. That model represents the global view of the server.
- Prediction Phase: Obtain outputs from each model in L for each data item in T and form an ensemble of training instances. The outputs of the nth instance t_n from the set of models in L are a set O_n: $O_n = \{o_{n1}, o_{n2}, ..., o_{nm}\}$. The set O consists of the ensemble of training instances for LCS $O = \{O_1, O_2, ...O_n\}$. The new server training set becomes $S = \{(O_1, c_1); (O_2, c_2); ...; (O_n, c_n)\}$.
- Learning Phase: Learning from data entries in the server training set S, $L* = LCS\{(O_1, c_1); (O_2, c_2); ...; (O_n, c_n)\}$.
- Output: descriptive model $L*$ obtained from the learning phase.

Majority Voting

An outcome of the server is contributed by all local models in DLCS. The server will choose the final outcome by voting between local models.

Assuming DLCS has a set L of m local LCS models $L = \{l_1, l_2, .., l_m\}$. For each testing instance $T_i \in T (T = \{(t_1, c_1); (t_2, c_2); ..; (t_n, c_n)\}$, where t is a training instance and c is an associated class) we have:

- Obtain predictive outputs from local models in L for each data item T_i.
- Voting is carried out between local predictive outputs in order to choose the most preferred outcome.
- The winning outcome is the one whose number of votes is greater than half of the numbers of individuals who are not indifferent between the two choices.

4.3 Traffic Load

In [8] we formed a formula for estimating the data transmission between clients and server for binary domain, based on the Minimum Description Length (MDL) principle in a similar way as [3]. The formula evaluates the complexity of the model in terms of data compression. The data needed for transmission between clients and the server includes the model and the training instances. Therefore, the cost of transmission is equivalent to the number of bits needed to encode the model (theory bits) plus training instances (exception bits).

$$MDL = TheoryBits(TL) + ExceptionBits(EL)$$

The length of the theory bits (TL) is the number of bits required to encode a set of classifiers that will be transferred in the network. In the previous papers, we ignored all parameters associated with each classifier. This section presents a formula that takes into account the length of all parameters associated with all classifiers.

The classifiers have a common structure: *Condition* $\longrightarrow$ *Action* : *parameters*. Their lengths are defined as follows:

$$TL = \sum_{i=1}^{nr}(TL_i) + \sum_{i=1}^{nr}(CL_i) + \sum_{i=1}^{nr}(PL_i)$$

Where nr is the number of classifiers of the model; TL_i, CL_i, PL_i are the length of the condition, action and parameters respectively. Assuming that the ternary representation (0,1,#) is used to encode the condition, and that each classifier has three parameters: fitness (a real value), numerosity (an integer number), and experience (an integer value), the length of each component of a classifier can be estimated as follows:

$$TL_i = nb$$

$$CL_i = log_2(nc)$$

$$PL_i = 2ni + nj$$

where nb is the number of bits required to encode a complete set of conditions, nc is the cardinality of the set of possible classes, ni is the number of bits required to encode an integer number, and nj is the number of bits required to encode a real number. For example, in the 20-multiplexer problem, $nc = 2$ while $nb = log_2(3^{20})$ since we have an alphabet of three symbols $\{0, 1, \#\}$ to encode 20 attributes corresponding to the 3^{20} instances. Therefore, the theory bits are estimated as:

$$TL = nr(nb + log_2(nc) + 2ni + nj)$$

Similarly, the exception part (EL) is estimated as follows:

$$EL = nu(na + log_2(nc))$$

where nu is the number of training instances sent to the server and na is the number of bits required to encode a training instance.

Thus, the length of data sent from a client to the server is:

$$MDL = (nr)(nb) + (nu)(na) + nr(2ni + nj) + (nr + nu)(log_2(nc))$$

5 Experimental Setup

We conducted several experiments with DLCS on the 20-bit multiplexer problem. The DLCS system is developed in C++. Unless it is stated otherwise, UCS in DLCS is setup with the same parameter values used by Wilson [25] and Bernado [4]: $v = 10, \theta_{GA} = 25, \chi = 0.8, \mu = 0.04, \theta_{del} = 20, \theta_{sub} = 20, N = 2{,}000$. Two points crossover and tournament selection with the size of 0.4 are used.

Each experiment is repeated in 10 runs. Each run uses different random seeds which are consistent in all experiments. The results reported in this paper are averaged over those 10 runs. Each iteration is a window of 50 training instances.

6 Combining Local Predictions at the Server

This section is designed to answer our first research question: do different methods of knowledge combination at the server affect the learning?

In previous papers, we combined the local predictions at the server using the knowledge probing approach and obtained promising results. In this section, we will compare the knowledge probing approach with one of the most simple and popular methods in ensemble learning: majority voting. All experiments in this section are set up with three clients.

6.1 Noise Free Environments

Before comparing the two approaches, we first investigate one of the most sensitive parameters of the knowledge probing approach: the training size of UCS at the server. The data is sent from clients along with their models for training the UCS on the top of local models. More training instances at the server would result in faster learning at the server.

Figure 2 presents the learning curves at the server with different training sizes. The training size refers to the amount of training instances that each client send to the server along with its update model. The data is chosen randomly from incoming instances at clients.

Clearly, learning is slowest with the training size of 50 instances. The difference is mainly observed during the first 300 iterations. Since all UCS in our system started without prior knowledge, they all require a period of time to build up their knowledge bases. Sending more training data to the server would provide more information and therefore speed up the initial learning of UCS.

Although learning with 1,000 instances is faster than with 500 instances, the difference fairly small. The learning curves of both training sizes are merged after 100 iterations. However, the training size of 1,000 instances requires the transmission of double the amount of training data. We decided to choose the training size of 500 instances for investigation from this point in the chapter.

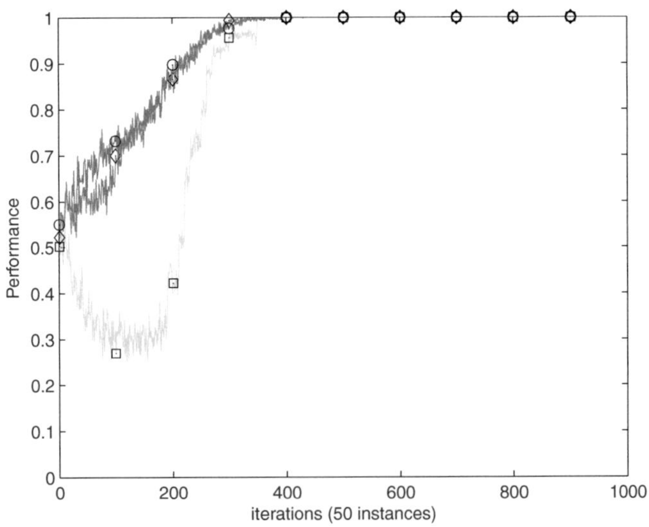

Fig. 2. The learning curves of the server. Experiments of the knowledge probing approach with different training sizes in noise-free environments – 50 instances (*square*), 500 instances (*diamonds*), 1,000 instances (*circle*)

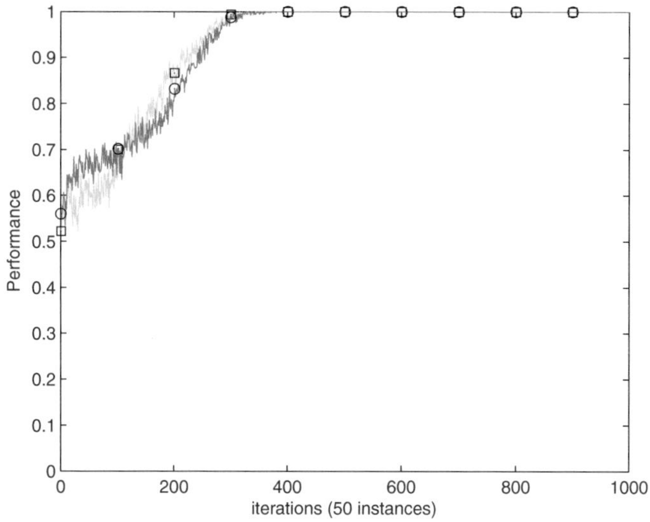

Fig. 3. The learning curves of the server. Experiments to compare the majority voting with the knowledge probing in noise-free environments – The knowledge probing approach (*square*), the voting approach (*circle*)

Figure 3 shows the learning curves at the server of DLCS using the voting and knowledge probing approaches. The learning at the server with the knowledge probing approach is slower at the beginning (the first 100 iterations) in comparison to the majority voting method. This is because the knowledge probing approach has another UCS on top of the local models. The server first forms its new training instances from the predictions of local models and then uses them for training the extra UCS. In order for UCS to obtain enough knowledge, initial training time is required. The majority vote, on the other hand, does not require learning time because the decision is based directly on voting between clients' predictions. However, once UCS accumulates enough knowledge in the knowledge probing approach, it starts to speed up its learning and converges faster than the voting approach. Both approaches are able to achieve 100% accuracy after about 300 iterations.

6.2 Noisy Environments

Noise is an unavoidable factor in the real world. This section investigates both methods in noisy environments, at two levels of noise: 0.1 (10 noisy instances in every 100 instances) and 0.2 (20 noisy instances in every 100 instances). Noise is added to the label of training instances by flipping their values between 0 and 1.

Figure 4 shows the learning curves at the server, using the knowledge probing and majority voting methods, with the noise level of 0.2. A similar pattern applies for noise level of 0.1.

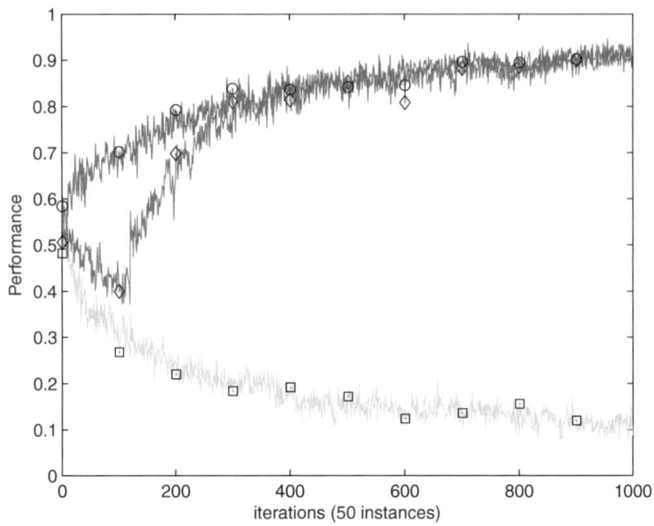

Fig. 4. The learning curves of the server. Experiments to compare the majority voting with the knowledge probing – The noise level is 0.2 – The knowledge probing approach with 50 instances (*square*), the knowledge probing approach with 500 instances (*diamond*), the voting approach (*circle*)

It is easily observed that the majority voting is more robust to noise than the knowledge probing as it learns faster and achieves higher accuracy than the knowledge probing approach in noisy environments.

In the knowledge probing approach, the training data of UCS at the server is constructed from predictions of local models. If predictions of local models are incorrect, the training data becomes noisy for training. At an early stage of learning, many predictions of local models are incorrect due to their inexperience since UCS normally requires a substantial time for the evolutionary and learning components to explore the search space and evolve the population. Once UCS receives enough training data, its knowledge (population of classifiers) becomes more reliable. There are also high chances of misclassifications by local models as the learning algorithm normally requires more time to eliminate the effects of noise. The knowledge accumulated in the system is unreliable due to noisy data encountered in the past and more training is required to reduce their impacts.

When the training size is too small (50 instances), the knowledge probing is not able to converge as with larger training size (500 instances). Clearly increasing the training size will help the system to overcome noise more quickly.

The voting method is more advantageous than the knowledge probing approach in noisy environment because it can reduce the bias in the system by their votes.

In conclusion, the knowledge probing approach learns faster than the voting approach on noise-free environments. When noise occurs in the data, the majority voting is more robust. Thus from this point, we will only consider the majority voting method for combining predictions at the server.

7 An Effect of the Number of Clients

This section targets to answer our second research question by investigating the predictive accuracy of the server with different number of clients.

7.1 Noise Free Environments

DLCS is simulated with 3, 5, and 9 clients in noise-free environments. Figure 5 shows the learning curves at the server with different numbers of clients.

The number of clients affects the predictive performance of the server only at the beginning. The server with 9 clients learn faster than with 3 and 5 clients. Increasing the number of clients from 5 to 9 results in a faster learning rate. At the beginning stage, the clients might not have yet discovered all hidden information due to bias in data distributions. Therefore having more clients gives more information to the server, and that results in faster learning by the server. Once the clients have learned completely, the performance at the server is not affected by the number of clients.

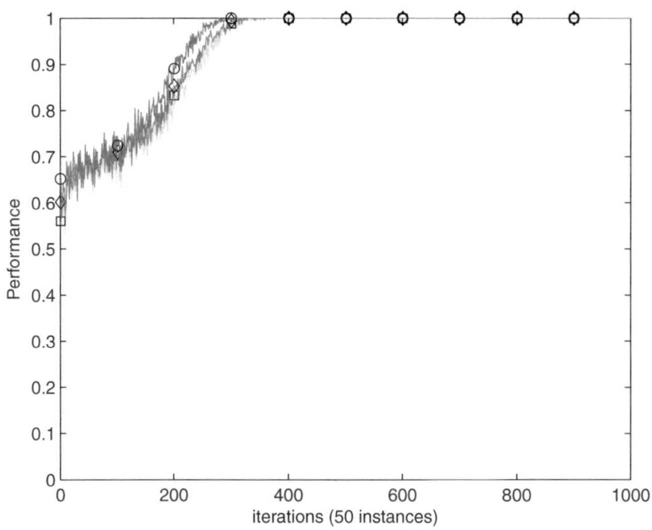

Fig. 5. The learning curves of the server. Experiments of DLCS with different numbers of clients in noise-free environments – 3 clients (*square*), 5 clients (*diamond*), 9 clients (*circle*)

7.2 Noisy Environments

Figures 6 and 7 show the learning curve of clients and server at a noise level of 0.2.

Figure 6 shows that when noise is added, clients take longer to learn and accuracy seems to be limited to around 85%. The population of classifiers ("NumMac" in Fig. 6) remains large: UCS is not able to evolve a compact population because of interferences from noise. Figure 7 shows that accuracy at the server is lower with 3 clients than with 5 or 9 clients. DLCS achieves an equivalent learning speed and accuracy with 5 clients or with 9. Hence adding more clients helps to increase the accuracy at the server, but only up to about 5 clients.

It is interesting to notice that the server is able to achieve around 10% better accuracy than the clients. This result is quite consistent with findings in the literature, as an ensemble of classifiers is often superior to the decisions made from any single learner.

In conclusion, the number of clients seems to affect the predictive accuracy of the server. More clients tends to result in better accuracy, but only up to a fairly small number of clients.

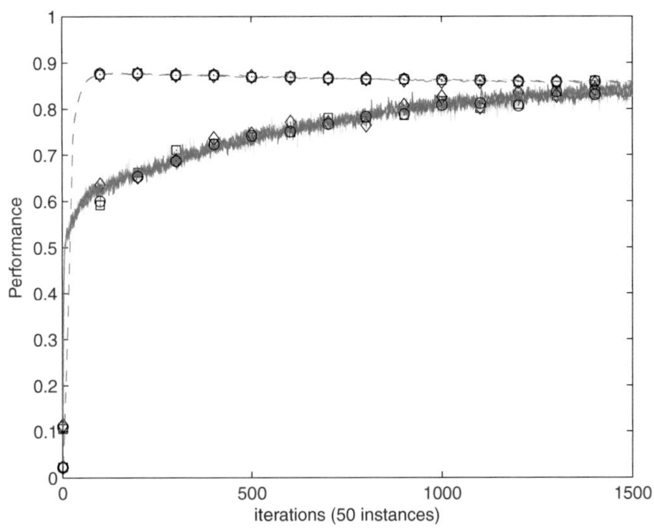

Fig. 6. The learning curves of clients. Experiments of DLCS with different numbers of clients – The noise level is 0.2 – The solid line represents the learning curve; the dash line represents the macro population, – 3 clients (*square*), 5 clients (*diamond*), 9 clients (*circle*)

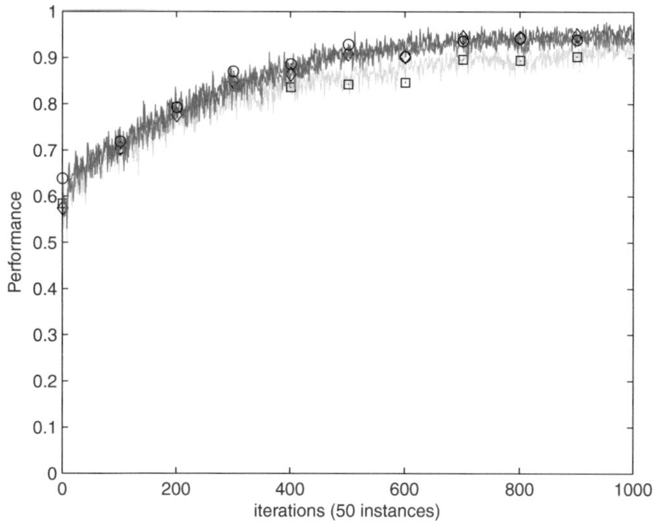

Fig. 7. The learning curves of the server - Experiments DLCS with different numbers of clients – The noise level is 0.2 – 3 clients (*square*), 5 clients (*diamond*), 9 clients (*circle*)

8 Investigation of Knowledge Passing in DLCS

This section is designed to understand the effects of the knowledge passing in DLCS, as stated in our third research question. DLCS builds up its knowledge over time by learning at clients. The knowledge of clients is represented by populations of classifiers. Knowledge passing refers to the migration of classifiers within the system. Two main channels of transmission are explored in this section: between the clients and the server, and between clients.

8.1 Between Clients and Server

The knowledge passed between clients and the server includes the clients' learnt models and some training data. The training data will be used for training at the server, if required in order to resolve any conflicts between local models. A collection of local models at the server represents the knowledge of the whole distributed databases. The training data is normally small in comparison to the local model. In particular, the simple voting approach does not require any training data at the server. Hence, the knowledge passing between clients and server in this section is mainly concerned with the transmission of local models without training data.

In our previous DXCS framework, populations of classifiers at clients are regularly transmitted to the server. The transmission data is reduced over time as the learnt models become more accurate and therefore compact. The advantage of this approach is that it is quite simple and also reserves the

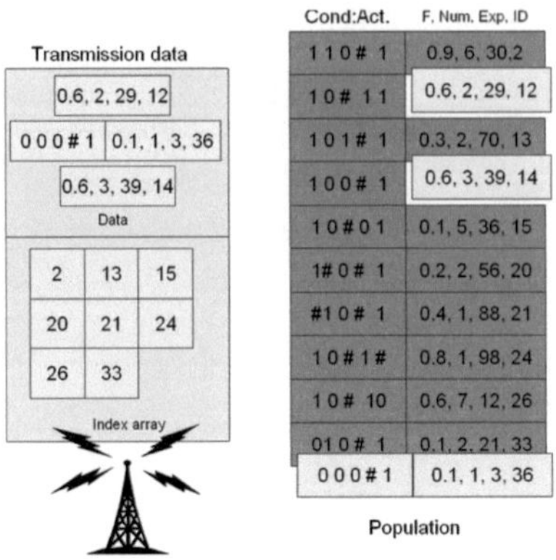

Fig. 8. The partial knowledge passing approach - An example of the data contained in a transmission packet from a client to the server

local knowledge at the server. However sending the whole population when the system is in a stable condition uses more network bandwidth than is actually needed.

To reduce the amount of data transmitted between clients and the server, in DLCS we introduce the approach of passing partial knowledge. In this approach, each classifier is associated with a distinct identification number (ID) that differentiates it from others in the population. The main purpose of this parameter is to synchronize the local models and their copies at the server without sending the whole population. Figure 8 illustrates the data transmission from a client to a server using the partial knowledge passing approach.

Data transmitted to the server consists of three main types of data: newly created classifiers, parameters of those classifiers that have been updated since the last transmission, and the unchanged index. The unchanged index contains the ID of those classifiers that have not been modified from the previous transmission. This index is used to synchronize the local model at the server.

When the data arrives at the server, it will have to wait for the server to update its corresponding local model. The server will first insert newly created classifiers into the population. The parameters of the updated classifiers are revised to record their new values. The server will discard classifiers that were not updated and whose ID was not listed in the unchanged index.

When we tested the approach using partial model transmission, the learning rate and accuracy were exactly the same as with full model transmission. This is as expected, since the difference between the approaches is just how the models are transmitted; the models still contain the same information.

Figure 9 shows the amount of data transmission from clients to the server, measured by the MDL equation, with full model and partial model transmission. It can be seen that the amount of data transmitted using the partial knowledge passing approach is about half that with full knowledge passing.

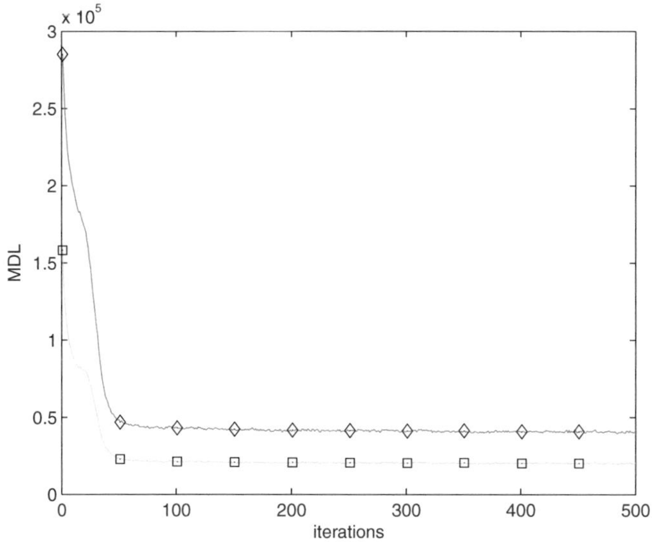

Fig. 9. The data transmission between clients and the server. Experiments to compare the partial with full models transmission approaches – The partial model transmission (*square*), the full model transmission (*diamond*)

Thus our partial knowledge passing helps to reduce substantially the network traffic while still maintaining equivalent accuracy and learning speed.

8.2 Between Clients

In this section, we focus on improving the learning speed at clients by sharing knowledge between clients. Bull et al. [5] presented the idea of rule sharing in ensembles of LCS. Their system consisted of 10 LCSs and was tested on the 20-bit multiplexer problem. The prediction of the system is computed by majority voting among the outputs from the 10 LCSs. The rule sharing takes place when the average time since the last rule sharing exceeds a threshold defined by users. A single rule is chosen according to fitness using the standard roulette-wheel selection. This rule is sent to another population in the ensemble. The sharing rules chosen from an action set found to be superior than from the whole population.

In our framework, we also apply this idea for rule sharing between clients. The rule sharing is carried out at the client level, where clients exchange their available knowledge to each other. The main difference with Bull's work is that our clients are trained with independent datasets and the migration classifiers are chosen in the population. The sharing decision is chosen after each fixed time window. Each client maintains its own temporary pools for keeping migration classifiers from other sites. After a time window, in our experiment we set to five instances, each client chooses a classifier from its population with probability proportional to its fitness. The client sends the chosen classifier to a random client in the system. The classifier is placed in the pool of that client waiting for integration into its population.

Figures 10 and 11 show the learning curves at the clients and server with and without the knowledge sharing. We can see that clients learn much faster with the knowledge sharing. UCS is able to achieve 100% accuracy after 400 iterations in comparison to 900 iterations without the knowledge sharing. The population with the knowledge sharing is able to cut down faster as it becomes stable after 400 iterations, which is around half time required without the knowledge sharing. The migration of classifiers in the system increase the diversity in the population that is why faster converge is observed.

However, its population is not able to achieve the same compactness as the population without the knowledge sharing. Continuing the knowledge sharing when the system is stable prevents UCS from compacting its knowledge because diversified classifiers are introduced to the population.

The improvement of learning speed at the clients has a direct impact at the server. The server is able to achieve 100% accuracy after 200 iterations in comparison to more than 300 iterations in the normal case.

Thus the knowledge sharing between clients helps to speed up the learning at clients and therefore at the server. However, the population size at is bigger with knowledge sharing than without. This is the trade-off between the faster learning speed and the compactness of the population.

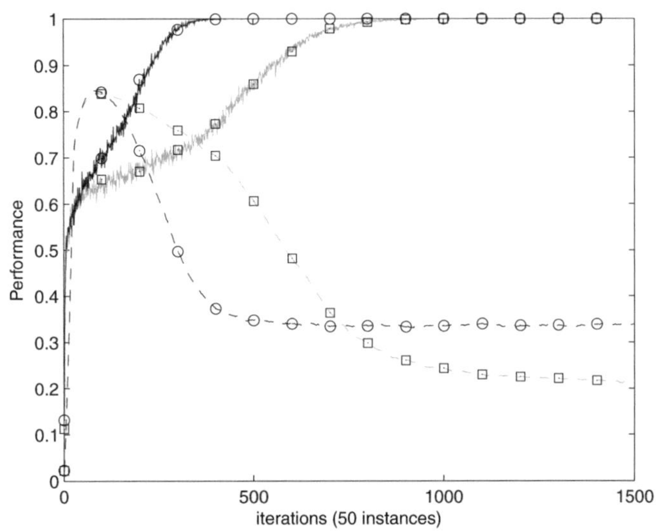

Fig. 10. The learning curves and population size of clients. Experiments to compare DLCS with and without knowledge sharing – The solid line represents the learning curve, the dash line represents the macro population – Without the knowledge sharing (*square*), with the knowledge sharing (*circle*)

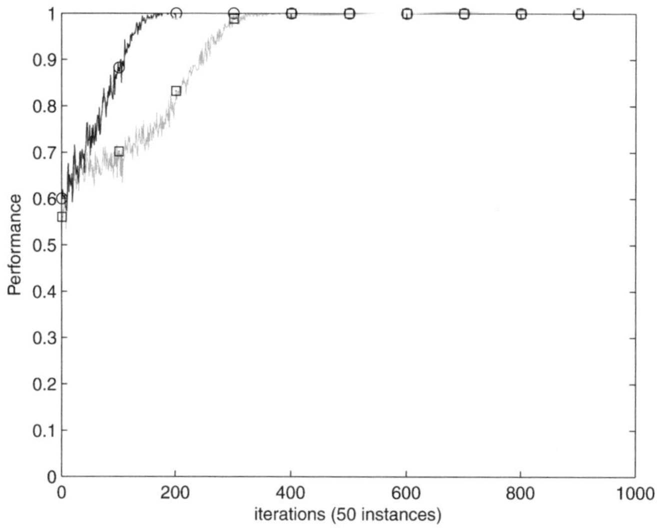

Fig. 11. The learning curves of the server. Experiments to compare DLCS with and without the knowledge sharing – Without the knowledge sharing (*square*), with the knowledge sharing (*circle*)

9 Conclusion

This paper investigates learning classifier systems in distributed computing environments. We proposed the distributed client-server framework called DLCS, where each client employs a complete learning classifier system that is trained on local data. The server stores local models in its memory and updates them whenever it receives new data from clients. The predictive decisions at the server are combined between predictions of those local models.

We tested DLCS on the 20-bit multiplexer problem in order to answer three main research questions.

The first question refers to the effect of different methods for combining predictions at the server. We compared the knowledge probing and majority voting methods. The result shows that different methods do affect the predictive performance at the server at the beginning. Knowledge probing produces faster learning than majority voting in noise-free environments. The majority voting is more robust than knowledge probing in noisy environments.

The second question relates to the number of clients in the distributed system. We hypothesized that increasing the number of clients would help the server to learn faster. The results show that increasing the number of clients from 3 to 5 provides more information to the server and therefore results in faster learning speed and better predictive accuracy at the server, but increasing the number of clients further has little effect.

The last question relates to knowledge passing within the system. We found that sending partial populations between client and server preserves learning at the server while reducing the traffic load in the system. Also, sharing knowledge among clients helps to speed up the learning of clients and also at the server. However, it has to be traded-off with the compactness of the population.

Acknowledgement

Work reported in this paper was funded by the Australian Research Council Linkage grant number LP0453657.

References

1. H. A. Abbass, J. Bacardit, M. V. Butz, and X. Llora. *Online Adaptation in Learning Classifier Systems: Stream Data Mining.* Illinois Genetic Algorithms Laboratory, University of Illinois at Urbana-Champaign, June 2004. IlliGAL Report No. 2004031.
2. J. Bacardit and M. V. Butz. *Data Mining in Learning Classifier Systems: Comparing XCS with GAssist.* Illinois Genetic Algorithms Laboratory, University of Illinois at Urbana-Champaign, June 2004. IlliGAL Report No. 2004030.
3. J. Bacardit and J. M. Garrell. Bloat control and generalization pressure using the minimum description length principle for a Pittsburgh approach learning classifier system. In *Sixth International Workshop on Learning Classifier Systems (IWLCS-2003)*, Chicago, July 2003.
4. E. Bernadó-Mansilla and J. M. Garrell-Guiu. Accuracy-based learning classifier systems: models, analysis and applications to classification tasks. *Evolutionary Computation*, 11(3):209–238, 2003.
5. L. Bull and T. Kovacs. *Foundations of Learning Classifier Systems: An Introduction*, volume 183, pp. 1–17. Springer, Berlin Heidelberg New York, June 2005.
6. P. K. Chan and S. J. Stolfo. Toward parallel and distributed learning by meta-learning. In *Working Notes AAAI Work. Knowledge Discovery in Databases*, pp. 227–240, Washington, DC, 1993.
7. V. Cho and B. Wuthrich. Distributed mining of classification rules. *Knowledge Information System*, 4(1):1–30, 2002.
8. H. H. Dam, H. A. Abbass, and C. Lokan. DXCS: an XCS system for distributed data mining. In *Proceedings of the Genetic and Evolutionary Computation Conference (GECCO-2005)*, Washington, 2005.
9. H. H. Dam, H. A. Abbass, and C. Lokan. Investigation on DXCS: An XCS system for distribution data mining, with continuous-valued inputs in static and dynamic environments. In *Proceedings of IEEE Cogress on Evolutionary Computation*, Edinburgh, Scotland, 2005.
10. T. G. Dietterich. Machine-learning research: Four current directions. *The AI Magazine*, 18(4):97–136, 1998.
11. U. Fayyad, G. Piatetsky-Shapiro, P. Smyth, and R. Uthurusamy. From data mining to knowledge discovery: An overview. In *Advances in Knowledge Discovery and Data Mining*, pp. 1–36. MIT, New York, 1996.

12. C. Giannella, R. Bhargava, and H. Kargupta. Multi-agent systems and distributed data mining. In *Cooperative Information Agents VIII: 8th International Workshop (CIA 2004)*, pp. 1–15, Erfurt, Germany, 2004.
13. D. E. Goldberg. *Genetic Algorithms in Search, Optimization, and Machine Learning.* Addision-Wesley, Reading, 1989.
14. Y. Guo, S. Rueger, J. Sutiwaraphun, and J. Forbes-Millott. Meta-learning for parallel data mining. In *Proceedings of the Seventh Parallel Computing Worksop*, 1997.
15. L. Hall, N. Chawla, and K. Bowyer. Combining decision trees learned in parallel. In *Proceedings of the Workshop on Distributed Data Mining of the 4th ACM SIGKDD International Conference on Knowledge Discovery and Data Mining*, 1997.
16. L. Hall, N. Chawla, and K. Bowyer. Decision tree learning on very large data sets. In *Proceedings of the IEEE Conference on Systems, Man and Cybernetics*, 1998.
17. J. H. Holland. Escaping Brittleness: The Possibilities of General-Purpose Learning Algorithms Applied to Parallel Rule-Based Systems. In Mitchell, Michalski, and Carbonell, editors, *Machine Learning, an Artificial Intelligence Approach. Volume II*, chapter 20, pp. 593–623. Morgan Kaufmann, Los Altos, 1986.
18. C. Jones, J. Hall, and J. Hale. Secure distributed database mining: Principles of design. In *Advances in Distributed and Parallel Knowledge Discovery*, pp. 277–294. MIT, 2000.
19. H. Kargupta and P. Chan, editors. *Advances in Distributed and Parallel Knowledge Discovery.* MIT, 2000.
20. T. Kovacs. Two views of classifier systems. In *Fourth International Workshop on Learning Classifier Systems - (IWLCS-2001)*, pp. 367–371, San Francisco, California, USA, 7, 2001.
21. A. L. Prodromidis, P. K. Chan, and S. J. Stolfo. Meta-learning in distributed data mining systems: Issues and approaches. In *Advances in Distributed and Parallel Knowledge Discovery*, pp. 81–114. MIT, New York, 2000.
22. S. F. Smith. *A Learning System Based on Genetic Adaptive Algorithms.* PhD thesis, University of Pittsburgh, 1980.
23. S. W. Wilson. Classifier fitness based on accuracy. *Evolutionary Computation*, 3(2):149–175, 1995.
24. S. W. Wilson. Generalization in the XCS classifier system. In J. R. Koza, W. Banzhaf, K. Chellapilla, K. Deb, M. Dorigo, D. B. Fogel, M. H. Garzon, D. E. Goldberg, H. Iba, and R. Riolo, editors, *Genetic Programming 1998: Proceedings of the Third Annual Conference*, pp. 665–674, University of Wisconsin, Madison, Wisconsin, USA, 1998. Morgan Kaufmann.
25. S. W. Wilson. Mining oblique data with XCS. In P. L. Lanzi, W. Stolzmann, and S. W. Wilson, editors, *Proceedings of the Third International Workshop (IWLCS-2000), Lecture Notes in Artificial Intelligence*, pp. 158–174, 2001.
26. D. H. Wolpert. Stacked generalization. *Neural Networks*, 5:241–259, 1992.

Knowledge Discovery from Medical Data: An Empirical Study with XCS

Faten Kharbat[1,2], Mohammed Odeh[1], and Larry Bull[1]

[1] School of Computer Science, University of the West of England, Bristol BS16 1QY, UK

[2] Computing Department, Zarqa Private University, Zarqa, Jordan, faten.west@gmail.com

Summary. In this chapter we describe the use of a modern learning classifier system to a data mining task. In particular, in collaboration with a medical specialist, we apply XCS to a primary breast cancer data set. Our results indicate more effective knowledge discovery than with C4.5.

1 Introduction

Learning Classifier Systems (LCS) [14] have been successfully used for data mining within a number of different medical domains, beginning with Bonelli and Parodi's [7] work using their 'Newboole' system. More recently, Wilson's XCS [41] was favorably compared to a number of other popular machine learning algorithms over numerous well-known benchmark data sets [5, 39]. XCS has been applied over the three Wisconsin Breast Cancer Datasets [6] and again achieved competitive results [1]. Other examples include a Newboole-like system, termed EpiCS [16], which was found to classify novel cases in another medical domain more effectively than decision rules derived from logistic regression [15].

This chapter presents a complete learning four-step discovery structure based on XCS using a breast cancer dataset obtained from a local health trust. This structure includes exploiting, compacting, and evaluating the generated knowledge in addition to assessment issues related to data preparation. It outlines the applicability of using XCS in a medical decision support task aimed at improving the diagnosis of primary breast cancer.

The chapter is organized into six sections: The four-phase knowledge discovery process is introduced in the next section with a brief description of each phase. The next four sections illustrate and discuss each phase separately. Finally, this practical investigation ends by presenting some of the learned lessons and conclusions.

F. Kharbat et al.: *Knowledge Discovery from Medical Data: An Empirical Study with XCS*, Studies in Computational Intelligence (SCI) **125**, 93–121 (2008)
www.springerlink.com

2 The Four-Phases Discovery Process

In an attempt to facilitate XCS along with diverse mechanisms to extract and discover knowledge from medical sources, a four-phase knowledge discovery process is outlined and illustrated in Fig. 1. Each phase is described briefly as follows.

Phase 1: Understanding the Dataset

In this phase, the initial data exploration is performed to verify the dataset completeness and missing attributes, to check the values of each attribute, and to examine the distribution of the existing diagnosis classes.

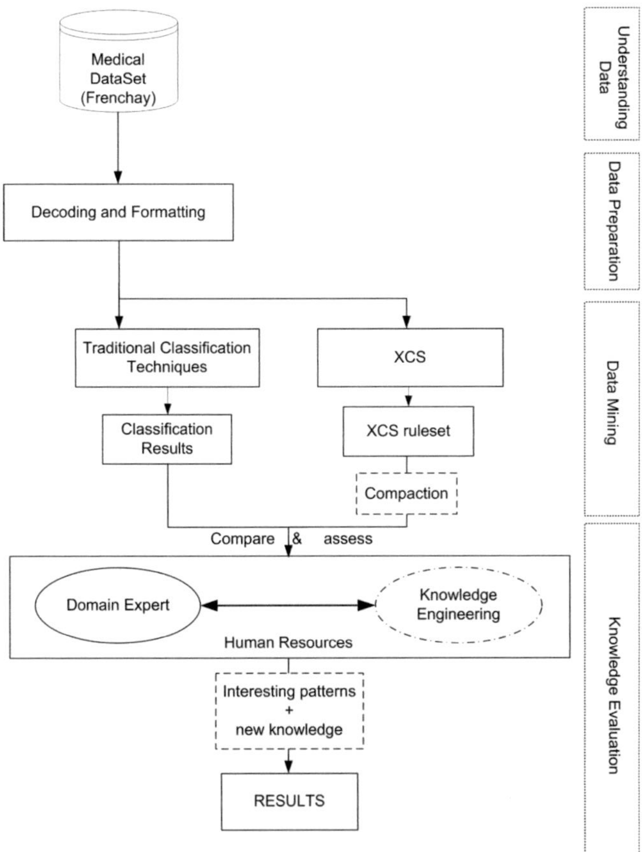

Fig. 1. Conceptual structure of the empirical investigation

Phase 2: Data Preparation

This phase is concerned primarily with the technical preparation of the dataset including the pre-processing and reformatting mechanisms to meet the requirements of the data mining techniques used within this investigation.

Phase 3: Data Mining and Knowledge Discovery

This phase is concerned with the extraction of patterns from the selected dataset over which XCS is applied in addition to other traditional classification techniques. Also, in this phase rule-driven compaction approach is introduced and applied over the ruleset generated from XCS.

Phase 4: Evaluating the Discovered Knowledge

This stage is concerned with evaluating and comparing the results obtained from XCS and the other used learning techniques to determine the extent to which they accurately classify pathological data. This evaluation is also carried out to assess the quality of rules generated from both XCS and the traditional learning techniques from the domain expert's point of view.

3 Phase 1: Understanding the Dataset

Although breast cancer research has been developing recently, the challenge has been to shift from gathering data to finding hidden patterns and trends that are most relevant to cancer diagnosis. Primary breast cancer diagnosis is a major challenge to oncologists who treat breast cancer since it is the first stage from where the cancer develops. Primary breast cancer refers to cancer that has not yet spread outside the breast. The Frenchay Hospital in Bristol, in the United Kingdom started building a database for primary breast cancer in 1999. Since then they have been developing their research studies and improving their treatment based on their outcomes and results. Thus, this investigation exploits one of their datasets which is complex and useful, as Frenchay hospital has been using it and relying on it since it contains accurate collected pathological data related to their patients.

The Frenchay Breast Cancer (FBC) dataset is a real-domain dataset which has a description of pathological data for women with primary breast cancer. The development of the FBC dataset started in 2002 by Dr. Mike Shere, a consultant pathologist in Frenchay Hospital, Bristol, UK.

Every case in the FBC is represented by 45 attributes, which collectively describe the status of breast cancer in a certain patient. Table 11 in Appendix illustrates the attribute names along with their ranges for numeric attributes. Tables 12 and 13 represent the binary and the categorical attributes, respectively.

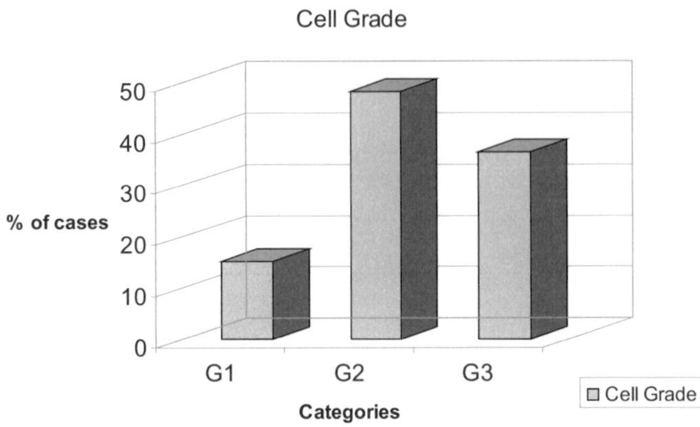

Fig. 2. Distribution of classes in FBC dataset

For the purpose of this investigation, 1,150 cases from the FBC dataset were used in this knowledge discovery process. The diagnosis for each case is the cell grade, which determines the aggressiveness of the breast cancer stage and has the three grades G1, G2, or G3, with the distribution of 15.2%, 48.3%, and 36.5%, respectively as shown in Fig. 2. The cell-grade is usually calculated by summing up some histological characteristics of breast carcinoma; and more specifically, it is the sum of the following attribute's values: Tubule-Formation-Score, Nuclear-Pleomorphism-Score, and Mitotic-Figure-Score.

4 Phase 2: Data Preparation

4.1 Data Pre-Processing

Data pre-processing takes place before applying machine learning techniques to solve some limitations and barriers found within the original data. This process should transform the original data into a more useful form [10]. In general, the problems within the original data extend from the existence of irrelevant attributes to the existence of multi-level noise, which prevents the knowledge discovery process from being successful.

Data pre-processing techniques vary, and there exist a huge number of techniques with different algorithms, as each rectifies a specific problem and suits certain machine learning techniques. Obviously, each algorithm has its strength and weakness points that may affect the original dataset. Filtering, noise modelling, feature selection, and data fusion are some of these techniques.

However, data pre-processing is a time consuming task given the need to (e.g., [10]): (1) determine what problems occur in the selected data, (2) determine the needed pre-processing techniques and select the best suitable algorithms for the used machine learning technique, (3) apply these over the original dataset to arrive at better resultant dataset. As this is not a closed or bounded problem, pre-processing the data is not treated here. This is because the aim of this investigation is to assess, evaluate, and compare the ability of XCS, along with other learning techniques, to classify and deal with raw data. Moreover, testing and evaluating each pre-processing technique with different learning techniques including LCS is beyond the scope of this research.

In the following sections, a simple preparation procedure is applied to setup the data in a suitable form. Formatting, decoding, and solving the imbalance problem within the FBC dataset were carried out in this phase.

4.2 Data Formatting and Decoding

Three types of attributes are used within the FBC dataset and these are: numeric, boolean and categorical. As in the related literature (e.g., [37]), numeric attributes are normalised between the values of 0 and 1, where a value X is decoded into X' using the minimum and maximum values of the attribute interval as follows:

$$X' = \frac{X - \min Val}{\max Val - \min Val} \tag{1}$$

The values in categorical and boolean attributes are decoded into values $\{0,\ 1,\ 2 \ldots n-1,\ \#\}$ where n is the number of the possible values of the attribute. Table 1 shows the decoding with a simple example of each attribute type.[1]

4.3 The Imbalance Problem

The class imbalance problem is a well-known problem, which occurs when the classes' representation is unequal in the dataset, and thus the classes' frequency is significantly unbalanced. The dominant class, which is represented more frequently within the dataset, is referred to as the majority class. Other classes which are represented by smaller sets are referred to as the minority classes. It is believed that this problem may hinder most of the learning algorithms from achieving high accuracy performance [2].

However, this problem seems to have relations with other real dataset problems. For example, Prati et al. [2004] revealed that class imbalance is not the only factor responsible for holding back the classification performance, but also affecting the degree of overlap between classes. Thus, solving the imbalance problem will not always increase the learning algorithm performance as there are other problems to be considered. Jo and Japkowicz [24] and Japkowicz [21] argued the same in that the imbalance problem may not be the main problem.

[1] XCS and the selected learning techniques are not affected by the ternary decoding used in the research. However, if other learning techniques are to be used (e.g., logistic regression) this may need some modifications.

Table 1. Attribute types: decoding and examples

Attribute type	No. of attributes	Decoding	Example
Numeric	15	Normalise between the values of 0 and 1	Attribute: age Min Max Original values 21 98 Decoding values 0 1 Example: 55 → 0.44
Boolean	13	Ternary	Attribute: DCIS-Necrosis Original values True False Any Decoding values 0 1 #
Categorical	17	0, 1, 2... $n-1$, #	Attribute: Core-Biopsy-B-code Values B3 B4 B4b Any Decoding 0 1 2 #

But they focused on the problem of the existence of small disjuncts[2] which correctly cover few data elements. These small disjuncts commonly are prone to higher error than large disjuncts. Several approaches were suggested to solve such problem in [21].

Although such studies have taken place, the other suggested problems (i.e., small disjuncts...etc.) do not have a clear simple solution, especially for a real-domain problem [21]. Therefore, dealing with them could just complicate, transform, or interpose the original dataset. Moreover, the imbalance problem still influences the performance of learning systems considerably; and therefore, this investigation will treat this problem as it may cause difficulties to learn concepts related to the minority classes.

There are two major categories of learning techniques designed to address the class imbalance problem [20]: supervised and unsupervised techniques. The supervised techniques have knowledge of the class labels, whereas the unsupervised techniques assume labels for the minority class [22], which is not the case in this research.

One category of supervised techniques that has been broadly used is related to the methods in which fractions of the minority and majority data elements are controlled via under-sampling and/or over-sampling, so that the desired class distribution is obtained in the training set [20]. Using the under-sampling technique, the data elements associated with the majority class are reduced [22]; and therefore, the size of the dataset is reduced significantly. Alternatively, over-sampling is usually applied to increase the number of data

[2] Small disjuncts are the learned classification rules that are generated from relatively few training examples [41].

elements of the minority classes [22], which will result in increasing the size of the dataset.

It was suggested in Barandela et al. [2004] that using over-sampling is required if the (majority/minority) ratio is very high. Moreover, Batista et al. [2] and Prati et al. [2004] suggested that the over-sampling method may generate more accurate results than under-sampling one. More precisely, random over-sampling showed competitive results than those more complex ones within their experiments. Wyatt et al.'s [45] results on XCS confirm the effectiveness of the simple over-sampling technique over more complicated ones.

For the FBC dataset, it can be seen that the class G2 is the major class and G1 is the most minor one with the ratio between G2 class and G1 (G2/G1) is 3.17 and the ration between G2 class and G3 (G2/G3) is 1.37; hence, balancing this dataset is required. After [45], the random over-sampling technique is chosen where random cases from the minority classes are selected and replicated so that all the classes in the dataset are represented by the same ratio.

4.4 Missing Data Problem

Missing data is another problem that occurs within real datasets. Within this research, missing data is treated while learning by XCS. In Holmes et al. [18] it is reported that XCS is stable across all the missing value types and densities which were illustrated in [17], and therefore this investigation does not analyse the density of the existing missing values or their types within the FBC dataset. However, it can be seen from Appendix that there are different types of missing values within the dataset which vary from simple uncollected elements to more complex missing values of the attributes. For example, the *specimen type* which has only five missing values is an example of simple missing values which may be attributed to unavailable data or just an error in data entry. This was referred to in [17] as missing completely at random. Another example is the *DSIC*, a dependant attribute that is *Histology*-based. If the *Histology* has the value of *M85203*, then the value of *DSIC* will not exist, else it will have a value. This is not a missing data that is not collected or corrupted in some way; but it is the nature of the attributes which have an entire dependency. However, in this investigation this is also considered as missing data.

Missing values are treated using the Wild-to-Wild mechanism [18]. While creating the match set [M], the matching process is performed in which the missing values in an input are handled by replacing them with don't care symbols or general intervals, and therefore they match any value. Covering an attribute is handled by assuming any missing attribute is a "don't care" or the most general interval for the attribute.

5 Phase 3: Data Mining and Knowledge Discovery

5.1 Well-Known Classification Techniques Results

Experiments were performed using the well-known and traditional classification techniques namely, Bayesian Network Classifier [23], Sequential Minimal Optimization [Platt, 1998], and C4.5 [35] using the Weka software [44]. These techniques were chosen for their performances over Frenchay dataset and because they are widely used within the AI community. All experiments were performed using the default parameters setting in Weka with the tenfold cross validation [28].

The Bayesian Network Classifier is a well known supervised learning technique and is one of the probabilistic, graph-based approaches to reasoning under uncertainty. It has shown competitive results in different cancer application domains such as the prediction of survival of patients with malignant skin melanoma [36], and the identification of 33 breast cancer risks [33]. In [32] the Bayesian Network was found to perform comparatively better than Neural Networks and logistic regression models in addition to its ability to explain the causal relationships among the variables. In the case of the FBC dataset, the accuracy performance for the Bayesian Network Classifier is $70.38\% \pm 5.15\%$[3].

The Sequential Minimal Optimization (SMO) is an optimization algorithm that quickly solves the Support Vector Machine (SVM) quadratic programming (QP) problem without any extra matrix storage and also without invoking an iterative numerical routine for each sub-problem [23]. Also, it has been used successfully in lung cancer to aid diagnosis [30], in addition to its competitive results for breast cancer diagnosis [29]. After applying SMO over the FBC dataset, the performance was found to be $72.50\% \pm 3.82\%$.

C4.5 is a well known decision tree induction learning technique which has been used heavily in the machine learning and data mining communities. The output of the algorithm is a decision tree, which can be represented as a set of symbolic rules. For the current dataset, results showed that the C4.5 technique achieved $77.4\% \pm 3.33\%$ classification accuracy with an average tree size of $101.51\% \pm 21.95\%$ and $70.51\% \pm 15.9\%$ as the number of obtained rules.

Table 2 shows a summary of the accuracy performance for the above three classification techniques. It can be seen that C4.5 achieved the highest performance among the selected classification techniques. Therefore, the generated rules (knowledge) from C4.5 are to be evaluated and compared with the rules obtained using compaction approaches over XCS solutions. This has been performed using the domain specialist to critically report on the results achieved as explained later in this chapter.

[3] Bayesian Classifier was tested and found it lacking compared to Bayesian Network Classifier. The accuracy performance for the Bayesian Classifier is $63.93\% \pm 4.64$.

Table 2. Accuracy performance for traditional classification techniques

Classification technique	Classification accuracy %	Tree size %	No. of rules %
Bayesian network classifier	70.38 ± 5.15	–	–
SMO	72.50 ± 3.82	–	–
C4.5	77.4 ± 3.33	101.51 ± 21.95	70.51 ± 15.9

5.2 XCS Results

This section shows the behaviour of XCS as described in [8] in learning the FBC dataset. Since the attributes in the dataset are divided into three data types: binary, categorical and real, the condition part of a rule in XCS combines real intervals, binary and categorical representations with their decoding as described above. Different representations were suggested in different problems. Initially, XCS used the ternary representation {0, 1, #} to encode the categorical condition part of a rule. Real attribute (Ai) in the condition is represented as an interval $[p_i,\ q_i]$, where p_i is the lower side of the interval and q_i is the upper side as described in [43]. For example, the following is the first seven interval predictors in the condition part in a rule in which the first and the seventh predictors use real intervals and the others use the categorical representation: (0.0–0.9)(#)(1)(#)(#)(3)(0.0–0.3)....

The Empirical Investigation

An empirical investigation was performed to determine a good setting for parameters, and was found that the classification performance is sensitive to the population size N, mutation step m_0, covering step r_0 and υ. As discussed Butz and Goldberg [9], a small population size hinders the solution to be generated because of the covering and reproduction processes. Different values of population size were tested (N = 5,000, N = 6,000, N = 8,000, N = 10,000, and N = 30,000). And, it has been found that the population size of N = 10,000 provides a sufficient population size as lower values did not allow an accurate solution to evolve.

The values of r_0 and m_0 determine the initial and intermediate movements in the solution map, where r_0 is the maximum covering step size for an attribute if no rule matches the current case; and m_0 is the maximum step size that an interval can widen to while in mutation. The effect of the value of υ has been illustrated in depth in [25, 26].

XCS was trained using tenfold cross validation for ten runs each over 1,000,000 iterations using roulette wheel selection with a population size of 10,000. The values for all other parameters are as follows[4]: $p_{\#} = 0.75$,

[4] A full description of the parameters and the XCS' algorithm is described in [8].

$\theta_{\mathrm{GA}} = 50$, uniform crossover $\chi = 0.8$, free mutation $\mu = 0.04$, $\alpha = 1$, $\delta = 0.1$, $\varepsilon_0 = 1.0$, $\theta_{\mathrm{sub}} = 20$, $\upsilon = 50$, $r_O = 0.4$, $m_O = 0.2$. Table 3 shows the classification accuracy and the average of the rulesets' size of XCS over the FBC dataset. It can be seen that XCS significantly outperforms C4.5 and the other traditional techniques in terms of its classification accuracy which is reassuring for the capability of XCS.

The Compaction Process

A three-phased approach (as depicted in Fig. 3) is proposed to combat the effects of the large numbers of XCS generated rules and to manage the complexity found among their condition elements. Also, in this proposed approach we aim to extract the hidden patterns which represent interesting knowledge, each of which describes a part of the problem space. The three stages in this approach are: pre-processing, clustering, and rule discovery which are explained in the next three sections respectively. In the first stage, rules' noise is reduced by eliminating weak rules, and the rules' encoding gets changed into a simpler binary representation to ease the infiltration into the condition element in order to extract representative patterns. In the second stage, the QT-Clust [13] algorithm is adapted and utilised to find the natural groups of rules from which the final stage extracts and discovers two types of patterns: compacted predictors and aggregate definite patterns.

Rule Pre-Processing

This phase takes place in two stages: filtering and encoding. First, the weak rules are identified and then removed from the original ruleset based on their

Table 3. Prediction accuracy of XCS over the Frenchay dataset

	Classification accuracy	Population size
Frenchay dataset	$80.1\%_{(5.9)}$	$7{,}974.4_{(157.4)}$

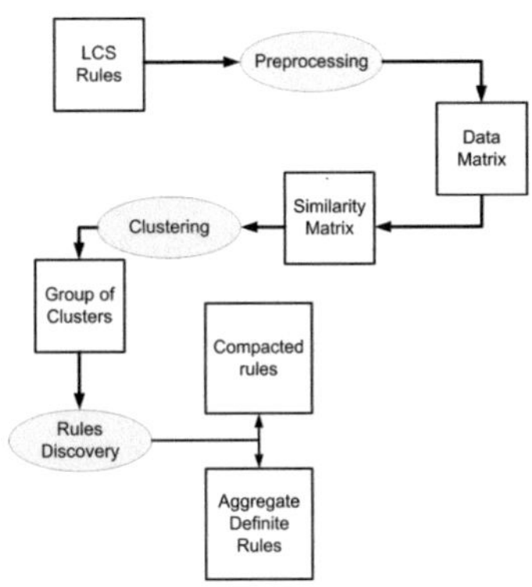

Fig. 3. The main stages of the proposed approach

low experience or *high prediction error*. The low experience of rules indicates that they either match a very small fraction of the dataset, which obviously could be matched by other rules, or by those that were generated late in the training process, which implies that XCS did not have enough time to decide whether to delete them or approve their fitness. Moreover, the high prediction error of a rule indicates its inaccuracy and/or that it has very significant missing information.

Filtering continues when the action of the rules that have sufficient experience with both *low estimated error* and *low prediction payoff* is converted to the opposite action. This step is reasonable in binary problems in which low prediction for a rule indicates that it belongs to the opposite action. However, this step should be modified later in the multi-action problem (i.e., with rules having three or more actions).

Rules with similar actions are isolated from other ones into a separated set since the purpose of the proposed approach is not to predict the action of the rule, but to find the natural groups having the same action. In the case of the binary action, there are only two sets: action one set and action zero set. In XCS this is not a straightforward step as the prediction array calculation plays a major role in the decision as to which action should be taken. This means that there could be some rules' conditions that exist in both sets. Therefore, it is essential to find these overlapped rules and place them in a separate set (overlapped action set) which can be considered to be ambiguous patterns that the expert needs to study further.

Rules Encoding

The second stage of this phase is the encoding stage. Encoding the rules signifies the transformation of XCS rules from the interval representation into the binary one. This step is only performed over the condition after the rules for each action were separated in the previous pre-processing stage. Each attribute in the condition is encoded using k binary bits where k is its possible discrete value [11]. The ith value ($i = 1..\ k$) is set to 1 if it is within the attribute range value, and 0 otherwise. For example, if a rule has two attributes (A_1 and A_2), each of which has ten possible values (1..10), and its conditions are:

$1 \leq A_1 \leq\ 4$ and $2 \leq A_2 \leq 5$ Rule$_{(1)}$

then each value has a representative bit which is equal to 1 if it is included within the condition, else it is 0. the binary encoding of the above condition is:

Attribute	A_1 (1–4)										A_2 (2–5)									
Possible value	1	2	3	4	5	6	7	8	9	10	1	2	3	4	5	6	7	8	9	10
(1) or (0)?	1	1	1	1	0	0	0	0	0	0	0	1	1	1	1	0	0	0	0	0

One of the advantages for this encoding is that it can support the mathematical and logical operations that can be executed over the rules. For example, extracting the overlapped cases between action-zero and action-one sets can be converted into the logical *and* (^) operator, as is the case between $\text{Rule}_{(1)}$ and the following rule ($\text{Rule}_{(2)}$):

$2 \leq A_1 \leq 3$ and $3 \leq A_2 \leq 7$ $\text{Rule}_{(2)}$

A_1	A_2	
0110000000	0011111000	$\text{Rule}_{(2)}$

Then, the newly formed overlapped rule is:

A_1	A_2	
0110000000	0011100000	$\text{Rule}_{(1)} \wedge \text{Rule}_{(2)}$

Note that if an attribute is not included into the rule then all its possible values will be set to "0", whereas if all the possible values were "1" then this attribute is a general one and can be considered as a don't care attribute "#". For example the following rule has not got the second attribute, and the first one is general:

$1 \leq A_1 \leq 10$ $\text{Rule}_{(3)}$

A_1	A_2	
1111111111	0000000000	$\text{Rule}_{(3)}$

Moreover, it is important to illustrate that if an attribute has all its bits equal to "0", then the whole rule is considered invalid and should be deleted. The output of this stage is the rules in the new binary representation which will feed into the next phase, namely the clustering phase.

Rule Clustering

This marks the second phase in this new approach where similar rules are identified and then grouped to form suitable clusters. Clustering is an unsupervised learning technique that is well-known for its ability to find natural groups from a given dataset, where data elements within one cluster have high similarities to each other but are dissimilar to other data elements in other clusters [12]. There are many clustering techniques in the literature, each with its own characteristics, limitations, and advantages. The categories of clustering techniques are "neither straightforward, nor canonical" [4]. However, studying the literature suggests that clustering algorithms can be classified into five main categories, namely: hierarchical methods, Self Organizing Map (SOM), partitioning, density-based, and grid-based methods. Although useful, these methods suffer from deficiencies which can be summarised as follows:

- Algorithms such as the SOM and the partitioning methods require the number of clusters needed as an input to drive the clustering process. Determining the number of clusters is also a well-studied area in the literature (e.g., [40]) and is still an open problem, and it seems "... impossible

to determine the right number of clusters" [46]. In order to adhere to the required number of clusters, a data element may be forced in or out of a cluster despite its low/high correlation with its other members, which can, in turn, decrease the quality and accuracy of the discovered clusters. Thus, changing such a parameter may considerably affect the final clustering result.
- The splitting and merging of clusters in the hierarchical methods, which are divided into agglomerative or divisive algorithms, is irreversible, which can lead to generating erroneous clusters.
- Density-based and grid-based methods are efficient in low-dimension space, but their efficiency may significantly be degraded if the dimension of the problem space is high [12]. Extensive approaches for feature selection have been long studied (e.g., [27, 31]) to reduce data dimensionality. However, using these techniques can result in loosing knowledge and critical data which will affect the accuracy of the produced clusters.

It is worth mentioning that the nature of XCS rules contributes and/or drives the selection, design, or adaptation of a certain clustering algorithm, simply for the following reasons. First, rules have the problem of "curse of dimensionality" [3]. That is, the dimension of the rule space (attributes) could be very high; and therefore, the selected algorithm should have the ability to deal with it in terms of the quality of the generated clusters and execution time. Therefore, the selected algorithm should capture the relevant part of the space and ignore the other irrelevant parts. Second, the singularity of some of the well-experienced rules that have low prediction error is an important issue since they may cover rare, unusual conditions, or dependencies between features. Therefore, these rules should not be forced into an existing cluster, but should be kept isolated in order to present them to the domain specialist to confirm their significance. This suggests that the selected algorithm should have the tendency to produce large clusters as well as small ones, where necessary.

For the above two reasons, the QT-Clust algorithm [13], which was introduced as a new paradigm of quality-based algorithms, has been adapted to fit the context of this research. This algorithm was originally developed for use with expression data to "... identify, group, and analyse coexpressed genes" [13]. Its main goal is to group similar elements into the largest possible clusters with their quality guaranteed. The algorithm is discussed in detail as follows.

Since the binary encoding has been used, $Rule_i$ is defined as a group of x bits as follows:

$$Rule_i(x) = \begin{Bmatrix} 0 \\ 1 \end{Bmatrix}$$
$$where\ x = 1, 2, \ldots, n$$

In our adaptation, the quality of a cluster is guaranteed by allowing $Rule_i$ to join a cluster C if it has *at least* similarity (σ_{QT}) with all the rules in the same cluster. Therefore, the quality of the cluster is quantified by its diameter which is defined within this work as:

$$\min_{i,j} \in C\{S_{i_j}\} \geq \sigma_{QT}$$

where i and j are rules in cluster C. S is the similarity measure being used to capture how related the two rules are. In our case, in XCS rules in the binary encoding, the Jaccard binary similarity coefficient [19] (S_{ij}) is used to measure the similarity between $Rule_i$ and $Rule_j$

$$S_{ij}(Rule_i, Rule_j) = \frac{p(Rule_i, Rule_j)}{p(Rule_i, Rule_j) + q(Rule_i, Rule_j)} * 100$$

where

$$p(Rule_i, Rule_j) = \sum_{x=1}^{n} Rule_i(x) \wedge Rule_j(x)$$

$$q(Rule_i, Rule_j) = \sum_{x=1}^{n} Rule_i(x) \oplus Rule_j(x)$$

$$and \,\{n = |Rule_i| = |Rule_j|\}$$

The Jaccard binary similarity coefficient has been used heavily in the related literature, and it means that the similarity measure considers only two situations: (1) when the two rules have a similar attribute value in which the x bit will be 1 (i.e., $p(Rule_i,\ Rule_j)$); and (2) when there is a difference between an attribute values, in which it is 1 in one rule and 0 in the other (i.e., $q(Rule_i,\ Rule_j)$). This similarity measure ignores the situation in which both rules have the attribute values equal to 0. For example, Jaccard similarity between $\text{Rule}_{(1)}$ and $\text{Rule}_{(2)}$ is explained below with p being in bold and q underlined:

A_1	A_2	
$\underline{1}\mathbf{11}\underline{1}000000$	$0\underline{1}\mathbf{111}\underline{00}000$	$\text{Rule}_{(1)}$
$\underline{0}\mathbf{11}\underline{0}000000$	$0\underline{0}\mathbf{111}\underline{11}000$	$\text{Rule}_{(2)}$

$$Sij_{(Rule_{(1)}, Rule_{(2)})} = \frac{\mathbf{5}}{\mathbf{5} + \underline{5}} * 100 = 50$$

This implies that there are five similar bits and five different ones. There are other possible similarity measures that can be used instead, but this simple measure was found to be efficient and suitable for XCS rules since the absence of an attribute value in both rules does not indicate a similarity but the existence does.

The Extended Jaccard similarity [38] (2) is applied on real and binary attributes which still shows the ratio of the shared attributes to the number of common ones.

$$sim_{ExJaccard}(x, y) = \frac{x \cdot y}{\|x\|_2^2 + \|y\|_2^2 - x \cdot y} \tag{2}$$

Where $\|x\|_2 = \sqrt{\sum_{i=1}^{n} |x_i|^2}$ which presents the L2-norm.

By calculating the similarity measure between all the rules, the similarity matrix is formed and passed to the QT-Clust algorithm to generate the possible clusters accordingly. The QT-Clust algorithm can be summarized as follows:

Input: the Similarity Matrix (SM) for all the rules. It assigns high score to similar rules which have considerable amount of similar attribute values, and low similarity value to others.

1. Do

1.1 maxLC = empty//maxLC represents the largest cluster that can be found from the remaining rules

1.2 For i = 1 to no of rules in SM

1.2.1 Open a new temporary cluster TC_i containing $Rule_i$.

1.2.2 Find all the rules in the SM that can join the cluster by maintaining its diameter.

1.2.3 If size(TC_i) > size(maxLC) then $\max LC = TC_i$

1.3 End for

1.4 Add maxLC to the final clusters set.

1.5 The rules within maxLC are removed from the SM so that they are not allowed to join another cluster.

2. Until all the rules are clustered (i.e., the SM is empty)

Output: the final group of clusters.

One of the issues that affects the quality of the generated clusters is choosing the value of σ_{QT}, as it could change the number and size of clusters. Increasing the value of σ_{QT} will not reduce the accuracy of the clusters since no unrelated rule will be forced to join a cluster. Whereas decreasing this value could ignore some differences between rules which could affect the cluster accuracy. Further discussion on the effect of σ_{QT} is illustrated in the next section.

Rule Discovery

The aim of clustering rules is to extract an explicit knowledge that identifies potential meaningful relationships between attributes in similar rules. Each cluster represents conjoint patterns describing part of the problem domain and revealing some evidence about complicated relations between its interior attributes. These discovered patterns could highlight implicit, previously unknown knowledge about the problem domain. However, the domain expert should be involved to sustain the benefits from these patterns and guide a future search for another.

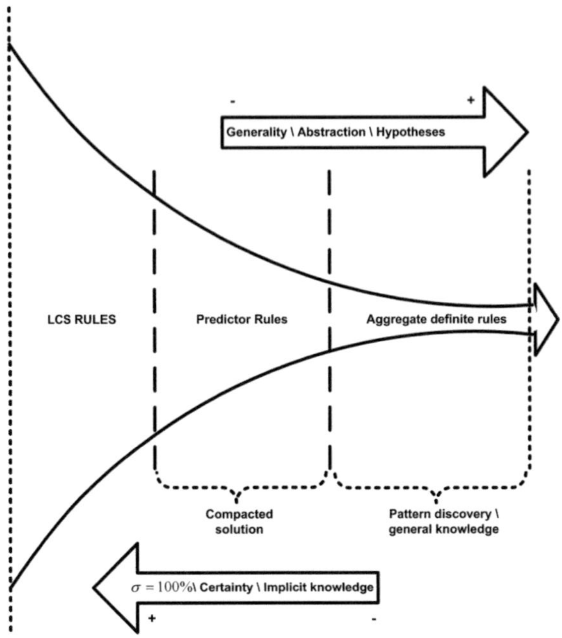

Fig. 4. An overview of rules' characteristics

Two levels of output are formed from each rule cluster generated from the previous stage. Each level represents an abstraction, generalization, and certainty level as shown in Fig. 4. First, in order to recast the implicit knowledge within LCS rules into a first level of explicit knowledge, an *Aggregate Average Rule* (AAR) of all the attribute values of all rules for each cluster is computed. The set of computed aggregate average rules are combined to form the *predictor ruleset* that can describe the original dataset. The following example explains these outputs visually.

In Table 4 an example of a cluster of ten rules (R1–R10) is described, where each rule has 15 attributes in the proposed binary encoding scheme. Having clustered these rules into one group, the existence of sufficient internal similarity between their attributes values has been revealed. This similarity drives the computation of the aggregate average rule of all the attribute values of the rules in this cluster so as to reveal the main characteristics of this cluster in the following computed aggregate average rule:

	A1	A2	A3	A4	A5	A6	A7	A8	A9	A10	A11	A12	A13	A14	A15
AAR	0	0	0	0	0		1	1	1	0	0	1	1	1	0

Hence, this rule represents the first level of transforming the rules in the cluster into a more general abstract and implicit knowledge. Furthermore, in

Table 4 some patterns have occurred within all the rules in the same cluster. For example, A7 has a value of "1" for all the ten rules in the same cluster to suggest that this attribute's value is a definite pattern within this cluster. In opposition to this pattern is the value of A6 which exists in most but not in all the rules of this cluster; and therefore; it is considered as an indefinite pattern. In order to reveal a high level of abstract general knowledge, the concept of the *Aggregated Definite Rule (ADR)* has been defined to represent the definite patterns that describe a common agreement among *all* attribute-value ranges to reveal the highly general characteristics for a given cluster. In this example, the aggregate definite patterns are summarised:

	A1	A2	A3	A4	A5	A6	A7	A8	A9	A10	A11	A12	A13	A14	A15
ADR	0	0	0	0	0	0	1	0	1	0	0	1	0	1	0

These aggregate definite patterns represent a generic, abstract, explicit type of knowledge which may be said to be in-line with "saying less and less about more and more" [34]. The quality of these patterns is mainly controlled by the similarity measure σ_{QT}, as increasing the value of σ_{QT} will increase the certainty of the patterns (see Fig. 4), since highly similar rules are grouped in clusters. On the other hand, decreasing the similarity level between clustered rules will widen the search area and enhance the discovery of more general patterns with lower level of certainty. Certainly, aggregate definite rules should be extracted within the clusters if sufficient number of participating rules exists. Table 5 shows the coverage accuracy and the size of the compacted solution after applying the rule driven approach (RDA). It should

Table 4. An example of a rule cluster

	A1	A2	A3	A4	A5	A6	A7	A8	A9	A10	A11	A12	A13	A14	A15
R1	0	0	0	0	0	1	1	1	1	1	1	1	1	1	0
R2	0	0	0	0	0	0	1	1	1	0	0	1	1	1	1
R3	0	0	0	0	0	0	1	0	1	0	0	1	1	1	0
R4	0	0	0	0	0	1	1	1	1	0	0	1	0	1	0
R5	0	0	0	0	0	1	1	1	1	0	0	1	1	1	0
R6	0	0	0	0	0	1	1	1	1	0	0	1	1	1	1
R7	0	0	0	0	0	1	1	1	1	1	0	1	1	1	0
R8	0	0	0	0	1	1	1	1	1	0	0	1	1	1	0
R9	0	0	0	0	0	0	1	0	1	1	1	1	1	1	1
R10	0	0	0	0	0	0	1	1	1	1	1	1	1	1	1

Table 5. Population size of the generated results

Compacted ruleset	Original population	Rule driven approach
Size	$7,974.4_{(157.4)}$	$341_{(27.16)}$
Coverage accuracy	$93.75\%_{(1..3)}$	$65.13\%_{(5.9)}$

be noted that comparison with the more traditional data driven approach to compaction, e.g., [44], highlighted a number of potential advantages to this new approach, such as less dependence upon the prediction array calculation scheme (see [25] for discussions).

6 Phase 4: Evaluation of the Discovered Knowledge

In general, the process of knowledge discovery aims to produce a novel piece of knowledge which can be easily comprehensible and useful in its problem domain. The significance of the generated knowledge can be assessed by the classification accuracy as discussed in the previous section. However, the value of the generated knowledge should also be evaluated from a domain expert point of view to emphasise if this knowledge fulfils the domain goals. That is, to evaluate the quality of the generated ruleset and to highlight the new piece of obtained knowledge (i.e., rules). This research concentrates on the medical domain, and breast cancer in particular. A medical expert, who is a consultant pathologist in the domain of primary breast cancer, has been involved in this evaluation study to report on the relative value of the extracted and compacted knowledge from the medical point of view. Figure 5 depicts a partial extract from the evaluation forms presented to the domain expert to assess the rules generated from the previous section:

The rules have been presented as a list of if-then statements and the domain expert was kindly requested to evaluate, relying on his past experience, the medical quality of each rule individually in terms of its *usefulness*, *contradiction*, and whether it is *interesting* and/or represents *new knowledge.*

RULES	*Usefulness* 1-5	*contradiction* Y/N	*Interesting/* *new knowledge* Y/N	*remarks*
. . .				
16. IF Immuno-ER-pos = TRUE and Mitotic-Figure-Score <= 1 and Histology = Ductal Carcinoma NST and Tubule-Formation-Score <= 7 THEN **Grade**=G1 (53)	1	N	N	
17. IF . . .				

Fig. 5. The assessment model

Usefulness means whether the rule would be of use in predicting or classifying future cases, with a scale of 1 (if the rule is of minor useful[5]) to 5 (if the rule is of the utmost significance and usefulness). *Contradiction* implies that the given rule contradicts with existing knowledge in the field or as per domain expert's background. A rule that is marked as *Interesting* and/or *new knowledge* indicates that this rule deserves future investigation since its diagnostic knowledge seems to be either not previously highlighted or that its existence maybe suspicious. Whenever a rule is masked as interesting, the domain expert is asked to provide a brief medical explanation to verify his point view in order to enrich the output assessment.

In this investigation each rule is associated with two numbers; the number of the correctly, and the incorrectly matched cases. These two numbers certify the rule's weight and accuracy. All the rules that have not had any matched cases are dropped out and not considered as efficient throughout this evaluation study.

6.1 Analysis of C4.5 Results

The randomly selected ruleset contains 85 rules in which the distribution of classes over the rules are 11%, 55% and 34% for G1, G2 and G3, respectively as illustrated in Fig. 6. Based on the expert's evaluation, six rules have been found to be of high usefulness, where five rules have been considered presenting new knowledge. However, none of the rules is found to be contradicting an existing knowledge. Tables 6 and 7 present the number of interesting rules and the rules' grade of usefulness for this group of rules, respectively.

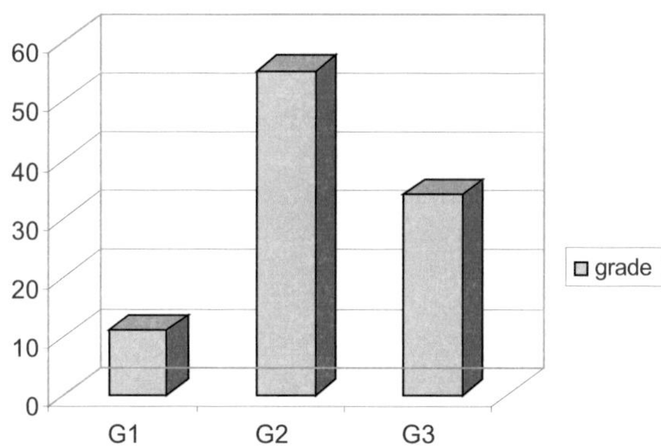

Fig. 6. The class distribution in the C4.5 ruleset

[5] Scale 1 is included into our empirical study; however, it is not shown in the results of this chapter since scale 1 represents the least useful rules and is beyond its motivation.

Table 6. Rules' distribution of grade of usefulness

Grade of usefulness	2	3	4	5
No. of rules (total number of rules = 85)	8	5	6	0

Table 7. Rules' distribution of correctness and new/interesting knowledge

	Yes
New/interesting knowledge	5
Contradicting	0

Table 8. Interesting rules from C4.5

No.	Condition	Grade
1	Immuno-ER-pos = FALSE	G2
	Immuno-Done = TRUE	
	LCIS-component = TRUE	
2	Immuno-ER-pos = FALSE	G3
	Immuno-Done = TRUE	
	LCIS-component = FALSE	
	Histology = Invasive Lobular Carcinoma	
	Report-Type = EX	
	Immuno-C-erb-B2-strength = Negative	
3	Immuno-ER-pos = FALSE	G3
	Immuno-Done = FALSE	
	DCIS-Necrosis = TRUE THEN Grade	
4	Immuno-ER-pos = FALSE	G3
	Immuno-Done = FALSE	
	DCIS-Necrosis = FALSE	
	Histology = Ductal Carcinoma NST	
	age <= 39	
5	Immuno-ER-pos = FALSE	G2
	Immuno-Done = FALSE	
	DCIS-Necrosis = FALSE	
	Histology = Ductal Carcinoma NST	
	age > 39	

Table 8 presents the interesting rules from C4.5 according to the domain expert's evaluation. The following notes summarise the evaluation of the domain expert of the generated C4.5 rules:

a. In general the generated rules from C4.5 were described by the expert as a simple, easy to understand, and useful.
b. C4.5 fails to discover some of the well-known primary cancer patterns such as the correlation between the number of involved nodes and the aggressiveness of the existing cancer.

c. Some rules were found too poor, maybe meaningless from the expert's point of view. For example, the $rule_i$ ($if\ sum > 5\ Then\ Grade = G3$), covers correctly about 95 cases (facts) and represents a naïve pattern.
d. It has been observed that the most useful rules in the expert's opinion match only few number of cases (facts) (i.e., between 3 and 10 cases), and that rules matched against a large number of facts seem to be not of a high value. That is, the over fitted rules seem to present a representative meaningful pattern, whereas the general rules describe useless patterns or even over general weak ones. The following rule is an example that covers more than 50 cases (facts) and has been considered as not useful at all.

   IF
      Immuno-ER-pos = TRUE
      and Mitotic-Figure-Score <= 1
      and Histology = Ductal Carcinoma NST
      and Tubule-Formation-Score <= 7
   THEN Grade = G1

   Alternatively, the following rule is considered to be of a high value (i.e., usefulness = 4) because it reveals the connection between this kind of histology and Grade 3 (G3) which can be easily used in predicting future cases, but this rule matches only three facts:

   IF
      Immuno-ER-pos = FALSE
      and Immuno-Done = TRUE
      and LCIS-component = FALSE
      and Histology = "Invasive Lobular Carcinoma"
      and Report-Type = EX
      and Immuni-C-erb-B2-strength = Negative
   THEN Grade = G3

e. Furthermore, none of the rules describing Grade 1 class (G1) were found interesting as all of them are considered as useless with their usefulness category between 1 and 2 (except one rule with usefulness = 3).

6.2 Analysis of XCS Results

A random ruleset was selected from XCS which contains a total of 2,901 rules and compacted using the proposed rule-driven compaction algorithms presented in Sect. 5.2.4. Figure 7 shows the percentage of rules' distribution over the three existing classes within the XCS ruleset before compaction.

It can be seen that this distribution is balanced and each class is roughly represented equally in the ruleset generated by XCS. After applying the rule-driven compaction algorithms, the size of the compacted rulesets was

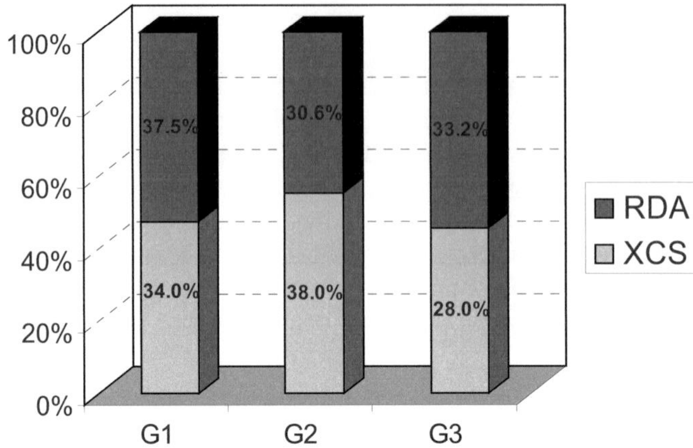

Fig. 7. Percentage of rules' distribution over the G1, G2, and G3 classes in XCS and rule driven approach to compaction in a randomly selected ruleset

Table 9. Rule-driven compaction: rules' distribution of grade of usefulness

Grade of usefulness	2	3	4
No. of rules (total number of rules = 300)	50	5	10

Table 10. Rule-driven compaction number of the correctness and new/interesting knowledge

	Yes
New/interesting knowledge	9
Contradicting	0

300. The compacted ruleset was then presented to the domain expert for evaluation.

For the rules compacted using the rule driven approach (300 rules), nine rules were described as interesting where ten rules were assigned to have a value 4 for usefulness. Tables 9 and 10 present the number of rules classified as useful and interesting using the rule-driven compaction, respectively. There were no contradicting rules in either results.

The following is the expert evaluation summary of the generated compacted sets categorised by the used compaction approach:

1. The expert reported that the compacted ruleset from the rule driven approach is able to find well-known patterns which described some parts of the problem domain. For example, some rules include the $Histology = Lobular\ Ca$ and describe Grade 2 (G2) diagnosis. This type of histology refers mostly to this kind of primary cancer. The ability to find such well-known patterns assures us of the power of XCS to extract representative

rules on the one hand and the rule-driven compaction approach in unveiling such hidden patterns on the other hand.

2. Very interesting relations and interactions were identified by the expert from the generated (ADRs[6]) and (AARs[11]) especially the ones related to grade 1 (G1) class. The expert reveals that ADRs and AARs present new knowledge regarding this type of primary cancer. Over 90% of the rules that were considered as interesting were related to Grade 1 (G1) diagnosis. In summary, these have shown that the *size of DCIS + invasive, Specimen type*, and *DCIS type* seem to be the most important attributes to determine this kind of primary breast cancer.
3. Some of the generated patterns were found inappropriate for use, as their left hand side consists of a high number of attributes. From the domain expert's point of view, it is very difficult to utilise rules having a large number of participating attributes in the real domain application. Therefore, this criterion may prevent many rules from being considered as interesting or useful. Moreover, the encoding and clustering technique led to some patterns with the following action: "*G1 or G2 or G3*" which represents a conflict pattern of symptoms and attributes. Such patterns are also not appropriate to be used in real applications.
4. Overall, the rules were judged to be more informative and qualitatively useful than the ruleset obtained from C4.5. They were found to contain more rich and descriptive patterns that cover the problem space more widely (see Appendix).
5. The expert revealed that because XCS generates larger rulesets, more time is needed to search for such unique, surprising, and newly hypothesised rules. This may be considered a barrier to arrive at the maximum benefit from the generated rulesets.
6. Although the classification accuracy of Grade 3 (G3) diagnoses is 91%, the expert does not find the rules covering this category interesting nor contributing to new knowledge. However, this may be attributed to the expert's experience being based on the cases collected from different patients and from his ongoing research. Grade 3 (G3) is the most aggressive primary cancer, on which more research is being carried out, more knowledge is evolving, and attention is given to provide a better understanding of this type of cancer, its causes and diagnosis. Therefore, although the generated rules cover over 90% of the cases (facts), it is suggested that the expert's expectation is higher than the knowledge hidden within the given cases (facts). Therefore, most of the generated rules that cover Grade 3 (G3) diagnosis were considered not interesting.

---

[6] Aggregate Definite Rules and Aggregate Average Rules were explained and discussed in Sect. 5.2.4.

7 Conclusion

This chapter has reported on the four-phase knowledge process applied to medical databases and, in particular, the area of primary breast cancer. Results indicate that using a learning classifier system, such as XCS, followed by the appropriate compaction process, can lead to knowledge discovery and rule induction from such datasets. The process started with preparing the data in terms of rule encoding and then balancing the representation of the involved data classes.

In addition, the dataset was applied to different learning techniques. XCS outperformed C4.5 and other selected learning techniques. The rulesets generated from XCS was compacted using rule driven approach and was qualitatively evaluated by a domain expert using the pre-defined criterion. The same criterion was used to evaluate the ruleset induced from C4.5 to allow the comparison to be carried out with rulesets from XCS.

In summary, rules evolved by XCS were qualitatively comparable to those induced from C4.5. Although C4.5 produces fewer rules, XCS was found to breed more useful descriptive ones. In the case of the C4.5 ruleset, the expert recognised some missing patterns and pointed out some poor quality rules. In XCS, rules are richer and representative but more complicated as well.

The investigation has been followed up to evaluate the rule driven approach which was able to extract and discover interesting patterns with their quality guaranteed. The extracted patterns consist of aggregate definite and aggregate average rules.

The aggregate definite rules revealed a high level of abstract general knowledge, revealing highly general characteristics of some parts of the current problem. However, they seem to include many attributes within the extracted patterns.

Applying XCS to a medical domain is not altogether new, but this chapter attains two achievements. First, the rules from XCS were compacted using a new algorithm which exploits rules to extract some of the hidden knowledge.

The involvement of a domain expert specialist who evaluated the rules relying on his previous knowledge guided by the assessment model enabled critical evaluation of the final rulesets to assess the quality of the rules generated from different perspectives. This involvement can be extended with more complicated assessment models to be utilised in refining the compaction technique or any related issue.

References

1. Bacardit, J., Butz, M. (2004). Data mining in learning classifier systems: comparing XCS with GAssis. In: *Advances in Learning Classifier Systems, 7th International Workshop, IWLCS 2004*, Seattle, USA, LNAI, Springer, Berlin Heidelberg New York

2. Batista, G., Prati, R., Monard, M. (2004). A study of the behaviour of several methods for balancing machine learning training data. SIGKDD Explorations, 6(1), 20–29
3. Bellman, R. (1961). Adaptive Control Processes: A Guided Tour, Princeton University Press, Princeton, NJ
4. Berkhin, P. (2002). Survey of clustering data mining techniques. Technical report, Accrue Software. Available from Accrue.com
5. Bernado, E., Llorà, X., Garrell, J. (2002). XCS and GALE: a comparative study of two learning classifier systems on data mining. In: *Advances in Learning Classifier Systems, 4th International Workshop*, Lecture Notes in Artificial Intelligence, vol. 2321. Springer, Berlin Heidelberg New York, pp. 115–132
6. Blake, C., Merz, C. (1998). UCI Repository of Machine Learning Databases. Irvine, CA: University of California, Department of Information and Computer Science. Available from http://www.ics.uci.edu/~mlearn/MLRepository.html, accessed 2/2004 [online]
7. Bonelli, P., Parodi, A. (1991). An efficient classifier system and its experimental comparison with two representative learning methods on three medical domains. In: *Proceedings of the 4th International Conference on Genetic Algorithms.* Morgan Kauffman, San Francisco, CA, pp. 288–295
8. Butz, M., Wilson, S.W. (2001). An algorithmic description of XCS. In: *Advances in Learning Classifier Systems, Proceedings of the Third International Conference–IWLCS2000.* Springer, Berlin Heidelberg New York, pp. 253–272
9. http://www.springerlink.com/content/5llgudt7wh6p37ay/
10. Famili, F., Shen, W., Weber, R., Simoudis, E. (1997). Data preprocessing and intelligent data analysis. *Intelligent Data Analysis*, 1(1–4), 3–23
11. Freitas, A. (2003). A survey of evolutionary algorithms for data mining and knowledge discovery. In: A. Ghosh, S. Tsutsui (eds). *Advances in Evolutionary Computing: Theory and Applications*, Natural Computing Series. Springer, Berlin Heidelberg New York, pp. 819–845
12. Han, J., Kamber, M., Tung, A. (2001). Spatial clustering methods in data mining: a survey. In: H. Miller, J. Han (eds). *Geographic Data Mining and Knowledge Discovery.* Taylor and Francis, London, pp. 188–217
13. Heyer, L., Kruglyak, S., Yooseph, S. (1999). Exploring expression data: identification and analysis of coexpressed genes. *Genome Research*, 9(11), 1106–1115
14. Holland, J.H. (1986). Escaping brittleness: the possibilities of general-purpose learning algorithms applied to parallel rule-based systems. In: R. Michalski, J. Carbonell, T. Mitchell (eds). *Machine Learning: An Artificial Intelligence Approach.* Morgan Kaufmann, San Francisco, CA
15. Holmes, J. (1997). Discovering risk of disease with a learning classifier system. In: T. Baeck (ed.). *Proceedings of the Seventh International Conference on Genetic Algorithms (ICGA97).* Morgan Kaufmann, San Francisco, CA
16. Holmes, J. (2000). Learning classifier systems: applied to knowledge discovery in clinical research databases. In: P. Lanzi, W. Stolzmann, S.W. Wilson (eds). *Learning Classifier Systems: From Foundations to Applications.* Springer, Berlin Heidelberg New York, pp. 243–261
17. Holmes J., Bilker W. (2002). The effect of missing data on learning classifier system: learning rate and classification performance. In: *Proceedings of 5th International Workshop, IWLCS*, Granada, Spain, September 7–8, pp. 46–60
18. Holmes, J., Sager, J., Bilker, W. (2004). A comparison of three methods for covering missing data in XCS. In: *7th International Workshop on Learning Classifier Systems (IWLCS-2004)*

19. Jaccard, P. (1912). The distribution of flora in the alpine zone. *The New Phytologist*, 11(2), 37–50
20. Japkowicz, N. (2000). The class imbalance problem: significance and strategies. In: *Proceedings of the 2000 International Conference on Artificial Intelligence (ICAI 2000)*, pp. 111–117
21. Japkowicz, N. (2003). Class imbalances: are we focusing on the right issue? Notes from the ICML Workshop on Learning from Imbalanced Data Sets II.
22. Japkowicz, N., Stephen, S. (2002) The class imbalance problem: a systematic study. *Intelligent Data Analysis*, 6(5), 429–450
23. Jensen, F. (1996). An Introduction to Bayesian Networks. Springer, Berlin Heidelberg New York
24. Jo, T., Japkowicz, N. (2004). Class Imbalances Versus Small Disjoints. *ACM SIGKDD Explorations Newsletter*, 6(1), 40–49
25. Kharbat, F. (2006). Learning Classifier Systems for Knowledge Discovery in Breast Cancer, Ph.D. Dissertation, UWE, UK
26. Kharbat, F., Bull, L., Odeh, M. (2005). Revisiting genetic selection in the XCS learning classifier system. In: *Proceedings of the IEEE Congress on Evolutionary Computation*, pp. 2061–2068
27. Kim, Y., Street, W., Menczer, F. (2003). Feature selection in data mining. In: J. Wang (ed.). *Data Mining: Opportunities and Challenges*. Idea Group Publishing, Hershery, PA, pp. 80–105
28. Kohavi, R., Provost, F. (1998). Glossary of terms. Editorial for the Special Issue on Applications of Machine Learning and the Knowledge Discovery Process. 30(2/3)
29. Land, W., Wong, L., McKee, D., Embrechts, M., Salih, R., Anderson, F. (2004). Applying support vector machines to breast cancer diagnosis using screen film mammogram data. In: *17th IEEE Symposium on Computer-Based Medical Systems (CBMS'04)*, pp. 224
30. Liu, W., Shen, P., Qu, Y., Xia, D. (2001). Fast algorithm of support vector machines in lung cancer diagnosis. In: *International Workshop on Medical Imaging and Augmented Reality*, June 10–12, Hong Kong, pp. 188–192
31. Molina, L., Belanche, L., Nebot, A. (2002). Feature selection algorithms: a survey and experimental evaluation. In: *Proceedings of the 2002 IEEE International Conference on Data Mining (ICDM'02)*, pp. 306–313
32. Moore, A., Hoang, A. (2002). A performance assessment of Bayesian networks as a predictor of breast cancer survival. In: *2nd International Workshop on Intelligent Systems Design and Application*, pp. 3–8
33. Ogunyemi, O., Chlebowski, R., Matloff, E., Schnabel, F., Orr, R., Col, N. (2004). Creating Bayesian network models for breast cancer risk prediction. In: *Cancer Risk Prediction Models, A Workshop on Development, Evaluation, and Application*, May 20–21, 2004, Washington DC, USA
34. Pantazi, S., Arocha, J., Moehr, J. (2004). Case-based medical informatics. *BCM Medical Informatics and decision making*, 4, 19
35. Quinlan, J. (1993). C 4.5: Programs for Machine Learning. Morgan Kaufmann, San Mateo, CA
36. Sierra, B., Larranaga, P. (1998). Predicting the survival in malignant skin melanoma using Bayesian networks: an empirical comparison between different approaches. *Artificial Intelligence in Medicine*, 14(1–2), 215–230
37. Sorace, J., Zhan, M. (2003). A data review and re-assessment of ovarian cancer serum proteomic profiling. *BMC Bioinformatics*, 4, 24

38. Strehl, A., Ghosh, J. (2000). Impact of similarity measures on web-pages clustering. In: *Proceedings of the 17th National Conference on AI (AAAI2000)*, pp. 58–64
39. Tan, K., Yu, Q., Heng, C., Lee, T. (2003). Evolutionary computing for knowledge discovery in medical diagnosis. *Artificial Intelligence in Medicine*, 27(2), 129–154
40. Tibshirani, R., Walther, G., Hastie, T. (2001). Estimating the number of clusters in a dataset via the gap statistic. *Journal of the Royal Statistical Society B*, 63, 411–423
41. Weiss, G. (2003). The Effect of Small Disjuncts and Class Distribution on Decision Tree Learning, Ph.D. Dissertation, Department of Computer Science, Rutgers University, New Brunswick
42. Wilson, S.W. (1995). Classifier Fitness Based on Accuracy. *Evolutionary Computation*, 3(2), 149–176
43. Wilson, S. (2000). Get Real! XCS with Continuous-Valued Inputs, Learning Classifier Systems. From Foundations to Applications. Springer, Berlin Heidelberg New York, pp. 209–222
44. Witten, I., Frank, E. (2005). Data Mining: Practical Machine Learning Tools and Techniques, 2nd edn. Morgan Kaufmann, San Francisco, CA
45. Wyatt, D., Bull, L., Parmee, I. (2003). Building Compact Rulesets for Describing Continuous-Valued Problem Spaces Using a Learning Classifier System. In: I. Parmee (ed.). *Adaptive Computing in Design and Manufacture VI*. Springer, pp. 235–248
46. Yeung, K., Haynor, D., Ruzzo, W. (2001). Validating clustering for gene expression data. *Bioinformatics*, 17, 309–318

Appendix

Table 11. FBC dataset: numeric attributes, their ranges, and missing values

No.	Attribute name	Missing value	Range
1	Grade	0	1–3
2	Age	0	21–98
3	Tumour-Weight	0	0–2,485
4	Tumour-Size	0	0–100
5	DCIS-extent-beyond-invasive-tumour	1	0–68
6	DCIS-Margin	756	0–30
7	Size-of-DCIS-Invasive	1	0–70
8	Nodes-examined-total	0	0–28
9	Involved-nodes-total	0	0–28
10	Immuno-pS2-score	228	0–200
11	Immuno-ER-H-Score	43	0–300
12	Immuno-PR-score	54	0–310
13	Tubule-Formation-Score	238	0–3
14	Nuclear-Pleomorphism-Score	238	0–3
15	Mitotic-Figure-Score	238	0–3

Table 12. FBC dataset: boolean attributes

No.	Attribute name	Missing value
16	Multifocal?	0
17	Multicentric?	0
18	DCIS-Necrosis	0
19	DCIS-component	0
20	Extra-Nodal-Invasion	0
21	LCIS-component	0
22	Immuno-Done	0
23	Immuno-pS2-pos	0
24	Immuno-ER-pos	0
25	Immuno-PR-pos	0
26	Immuno-C-erb-B2-pos	0
27	Immuno-Bcl-2-pos	0
28	Final-Operation	0

Table 13. The categorical attributes in Frenchay dataset with their possible values and distribution among the data

No.	Attribute name	Missing value	Possible values
29	Exscision-Margin-score	340	A,B,C,D,U
30	Core-Biopsy-B-code	890	B3,B4,B4b,B5,B5a,B5b,B5c
31	Primary site	0	C50.0, C50.1, C50.2, C50.3, C50.4, C50.5, C50.6, C50.9
32	Specimen-type	5	Completion-Mastectomy, Mastectomy, Core-Biopsy, Further-excision, Further-excision-and-nodes, Mammotome-Biopsy, Nodes-only, Skin-Biopsy, Skin-Punch-Biopsy, Surgical-Excision, Touch-Prep-from-Core
33	Involved-Margin-which	1,096	Deep, Inferior, Lateral, Medial, Superficial, Superior
34	DCIS-Nuclear-Grade	551	High, Intermediate, Low
35	latescality	0	L,R,B
36	Histology	0	M-82113M-85003, M-2105, M-72175, M-0103, M-80503, M-82113, M-4803, M-85002, M-85003, M85003-M85203, M85005, M5042, M85123, M85202, M5203, M-85205, M-85403, M-5413, M-85405, M85003
37	Histological-Calcification	376	Malignant, Benign, Absent, Benign-and-Malignant
38	Synchronous-Tumour	276	N,Y
39	Lympho-vascsclar-invasion	348	N,Y,X,U
40	Immuno-Bcl-2-strength	158	Negative, Strong, Mod, Weak
41	Immuno-C-erb-B2-strength	74	Negative, Strong-3, Weak-1, Mod-2

Table 13. (continued)

No.	Attribute name	Missing value	Possible values
42	Pathological-M-Stage	347	pM0,pMX
43	Pathological-N-Stage	341	pN0,pN0-i, pN1a, pN1bi, pN1bii, pN1biii, pN1biv, pN1mi,pN2, pNX, pN2a,pN3, pN3a, pN3b
44	Pathological-T-Stage	340	pT1a, pT1b, pT1c, pT1mic, pT0, pT2, pT3, pT4b, pTis, pTis-DCIS, pTis-Paget, pTX, ypT0, ypT1a, ypT1c, ypT2, yT0
45	DCIS-type	667	Solid-and-Micro-papillary, Solid-cribriform-Comedo, Solid-cribriform, Cribriform-Comedo, Comedo,Cribriform, Solid-and-Comedo, Solid-Comedo,Comedo-Solid, Cribriform-and-Comedo, cribriform-and-Solid, Micro-papillary, Micro-papillary-and-Comedo, Micro-papillary-and-Cribriform, Papillary,Solid, Solid-and-Cribriform

Mining Imbalanced Data with Learning Classifier Systems

Albert Orriols-Puig and Ester Bernadó-Mansilla

Enginyeria i Arquitectura La Salle, Universitat Ramon Llull, Quatre Camins, 2. 08022, Barcelona, Spain, aorriols@salleURL.edu, esterb@salleURL.edu

Summary. This chapter investigates the capabilities of XCS for mining imbalanced datasets. Initial experiments show that, for moderate and high class imbalances, XCS tends to evolve a large proportion of overgeneral classifiers. Theoretical analyses are developed, deriving an imbalance bound up to which XCS should be able to differentiate between accurate and overgeneral classifiers. Some relevant parameters that have to be properly configured to satisfy the bound for high class imbalances are detected. Configuration guidelines are provided, and an algorithm that automatically tunes these XCS's parameters is presented. Finally, XCS is tested on a large set of real-world problems, appearing to be highly competitive to some of the most well-known machine learning techniques.

1 Introduction

Learning Classifier Systems (LCSs) [13] are complex machine learning methods that combine reinforcement learning and evolutionary computation techniques to evolve novel knowledge. Initially designed regarding the animal learning and cognitive psychology, some LCSs, and specially XCS [32, 33], have been adapted to solve ambitious machine learning problems, facing new challenges. Among them, learning from datasets that contain *rare objects* has been identified as one of the biggest challenges to many well-known learners in the machine learning realm. Many examples of real-world domains that contain rare objects can be found, such as fraudulent credit card transactions [11], learning word pronunciation [12], and detection of oil spills from satellite images [21].

Research on the detection of rare objects has been conducted from two different perspectives. The first perspective, associated mainly to supervised tasks, focuses on the drawbacks caused by training the learner with datasets that contain different proportion of instances per class. As many learners tend to minimize a global measure of error, they might be biased toward the most numerous classes in the training dataset [19,20]. The second perspective, associated to both non-supervised and supervised tasks, is concerned about the

A. Orriols-Puig and E. Bernadó-Mansilla: *Mining Imbalanced Data with Learning Classifier Systems*, Studies in Computational Intelligence (SCI) **125**, 123–145 (2008)
www.springerlink.com

distribution of examples around the feature space. It analyzes the difficulties of learning from datasets that contain rare cases around the feature space, which cause *small disjuncts* [17]. Both perspectives are really close, since imbalanced datasets usually present rare cases. In fact, rare cases or small disjuncts have been addressed as *within-class imbalances* [18]. In this paper we analyze the effect of class imbalances as a whole, but also the difficulties induced in learning classifiers systems due to *small disjuncts*.

Some studies have analyzed the behavior of LCSs on imbalanced datasets, and some approaches have been proposed to boost their learning capabilities on imbalanced data. Holmes addressed this topic in the context of epidemiological data, enhanced EpiCS [14] with a strategy based on disproportionate reinforcement per class [15], and lately built EpiXCS [16], an XCS-like system derived from EpiCS. Other studies [22, 23] empirically analyzed the effects of imbalanced data on UCS [3], a learning classifier system derived from XCS specialized for classification tasks. Results evidenced that UCS was biased towards the majority class in highly imbalanced domains, and resampling techniques were presented as effective methods to alleviate the problem. Not until recently the first analyses of XCS's performance on imbalanced data have been done [24], showing that XCS is quite robust to class imbalances if it is properly configured.

The understanding of a complex system as XCS is crucial before applying it to solve real-world tasks. Aiming at this objective, this paper extends the study of XCS on imbalanced domains made in [24]. We provide a bound on the maximum amount of imbalance with which XCS can correctly deal, and we identify relevant XCS's parameters that need to be set properly to satisfy this bound in highly imbalanced datasets. Guidelines on how to tune these parameters are provided, and new experiments reveal that an appropriate setting of XCS's parameters drastically improves XCS's performance for high class imbalances. However, configuration guidelines depend on characteristics that may be unknown for real-world problems. Therefore, we present a method that automatically configures XCS's parameters, based on information collected during learning. The conclusions obtained from the analysis of XCS in artificial imbalanced domains are then applied to XCS so that it can efficiently mine imbalanced data from real-world datasets. Specifically, the algorithm of XCS's self-adaptation gives competitive results compared to learners such as C4.5, which is particularly known for yielding good performance in imbalanced datasets, as well as SMO and IBk.

The remainder of this chapter is organized as follows. Section 2 briefly introduces XCS for data mining. Next, we test XCS on an artificially imbalanced domain (Sect. 4). We model classifier's error, identify XCS's relevant parameters and provide guidelines on how to configure them in Sect. 4. Then, XCS is compared to other highly-competent learners on a large set of real-world problems. Finally, Sect. 7 summarizes, concludes and discusses further work.

2 Description of XCS

XCS is an online accuracy-based LCS that solves a problem by evolving a set of sub-solutions distributed in *niches* around the problem space. This section provides a brief description of XCS restricted to classification tasks. For further details, the reader is referred to [32, 33]; besides, an algorithmic description can be found in [10].

2.1 Knowledge Representation

XCS evolves a population of classifiers [P], where each classifier consists of a production rule of the form *condition* → *action* and a set of parameters. The most important parameters are: (a) the prediction p, which estimates the payoff that the classifier will receive when the rule is fired and its action is chosen as the output, (b) the prediction error ϵ, which estimates the error between the prediction and the received payoff, (c) the fitness F, which evaluates the accuracy of the classifier with respect to other classifiers in the same *action set*, and (d) the numerosity *num*, which counts the number of copies of the classifier in the population.

The condition representation was originally represented in the ternary alphabet: $\{0, 1, \#\}^\ell$, where ℓ is the length of the input instance. The symbol #, called *don't care*, allows to express generalizations in the classifier's condition. For real attributes, the input is codified as a set of intervals $[l_i, u_i]^\ell$, where l_i and u_i represent the lower and upper values that the attribute can take to apply the rule.

2.2 Performance Component

At each time step t, an input instance s_t is sampled, and XCS builds the *match set* [M], which contains all the classifiers in [P] that match the input instance. If the classifiers in [M] predict less that θ_{mna} different actions, the *covering* operator is activated, which creates new classifiers with a condition generalized from s_t and an action chosen from the ones not represented in [M] until θ_{mna} different actions are covered. For each action a_i in [M], XCS computes the payoff prediction $P(s_t, a_i)$ as a fitness weighted average of the prediction of all classifiers in [M] advocating a_i. Then, XCS selects an action to perform. Different selection regimes can be applied: from a *pure-explore regime*, in which the action is randomly selected, to a *pure-exploit regime*, in which the action with the highest prediction is chosen. Under classification problems, typically, the pure-explore regime is used during training, while the pure-exploit regime is used when the system predicts new unknown instances. Finally, XCS creates the action set [A], consisting of all classifiers in [M] that predict the chosen action.

2.3 Parameter's Update

The chosen action is sent to the environment, which returns a reward R that is used by XCS to update the parameters of the classifiers in [A]. First, the prediction p is adjusted: $p = p + \beta(R - p)$, where β is the learning rate $(0 < \beta \leq 1)$. Next, the error ϵ is updated: $\epsilon = \epsilon + \beta(|R - p| - \epsilon)$. To update the classifier's fitness, XCS first computes the *accuracy* κ as follows:

$$\kappa = \begin{cases} 1 & \text{if } \varepsilon < \varepsilon_0 \\ \alpha(\varepsilon/\varepsilon_0)^{-\nu} & \text{otherwise} \end{cases} \tag{1}$$

where ϵ_0 is the maximum error that a classifier can take to be considered accurate, and α and ν are constants that control the rate of decline in accuracy $(0 < \alpha \leq 1$ and $\nu > 0)$. κ is used to compute the *relative accuracy* κ' of the classifier in [A]: $\kappa' = \kappa / \sum_{[A]} \kappa_i$. Finally, the fitness is updated from the relative accuracy: $F = F + \beta(\kappa' - F)$. Note that the fitness is shared among the classifiers in the same action set since it is calculated from the relative accuracies.

2.4 Discovery Component

In XCS, the *genetic algorithm* (GA) is applied to the action set with a frequency fixed by the parameter θ_{GA}. It selects two parents from [A] (following either a proportionate selection scheme [32] or a tournament selection scheme [9]) and copies them creating two new classifiers, which are crossed and mutated with probabilities χ and μ respectively. The resulting offspring are introduced in the population, applying subsumption if required [33], and two classifiers are deleted to keep the population size constant.

3 XCS and Class Imbalances

In this section, we investigate how different amounts of class imbalance affect XCS. For this purpose, we designed a problem that permits to vary the complexity along the imbalance dimension: the *imbalanced multiplexer* [24].

3.1 The Imbalanced Multiplexer

The *multiplexer* [32] is a completely-balanced problem defined for binary strings of size ℓ, where the first $\log_2 \ell$ bits are the *address bits* and the remaining bits are the *position bits*. The output is the value of the position bit indicated by the decimal value of the address bits. The *imbalanced multiplexer* [24] permits to control the imbalance complexity of the multiplexer by undersampling the class labeled as '1'. In the remainder of this paper, we use the *imbalance ratio* – i.e., the ratio between the number of instances sampled of the majority and the minority class – to refer to the imbalance complexity.

3.2 XCS on the Imbalanced Multiplexer

We ran XCS on the 11-bit multiplexer with the following configuration[1]: N = 800, $\beta = 0.2$, $\theta_{GA} = 25$, $\epsilon_0 = 1$, $\alpha = 0.1$, $\nu = 5$, $\chi = 0.8$, $\mu = 0.04$, $\theta_{del} = 20$, $\delta = 0.1$, $\theta_{sub} = 200$, $P_{\#} = 0.6$. We used roulette wheel selection, two-point crossover, and niched mutation [10] for the genetic algorithm. Subsumption was applied in the GA, but not in the action set. Figure 1 shows the percentage of correct classifications of the majority class (TN rate) and the percentage of correct classifications of the minority class (TP rate) obtained by XCS with imbalance ratios from $ir = 1$ to $ir = 512$. Curves are averaged over ten runs. Note that, for $ir = 1$, the same proportion of instances per class is sampled; on the other hand, for $ir = 512$, there are 512 instances of the majority class sampled for each instance of the minority class.

Figure 1a, shows that the TN rate quickly raises to 100% for any imbalance ratio tested. On the other hand, Fig. 1b illustrates that XCS only achieves 100% TP rate for $ir \leq 16$. For $ir = 32$, the TP rate is about 95% after 10^6 iterations, reaching 100% after $2 \cdot 10^6$ explore trials (not shown in the graph for a better visibility). However, for higher imbalance ratios, XCS cannot achieve 100% of TP rate. For $ir = 64$, the TP rate remains below 20%, and increasing the number of learning iterations does not provide any improvement. For $ir > 64$, the system classifies all the inputs as they belonged to the majority class.

To explain the degradation of the TP rate with ir, we checked the final populations for the different imbalance ratios. For the lowest imbalance ratios ($ir \leq 16$), the final populations contained few *overgeneral classifiers* – classifiers that match training instances of different classes –, all of them with high error. For $ir > 16$, the numerosity of overgeneral classifiers increased exponentially with the imbalance ratio. For $ir = 64$, overgeneral rules represented about 15% of the final population; for $ir = 128$ (see Table 1), they

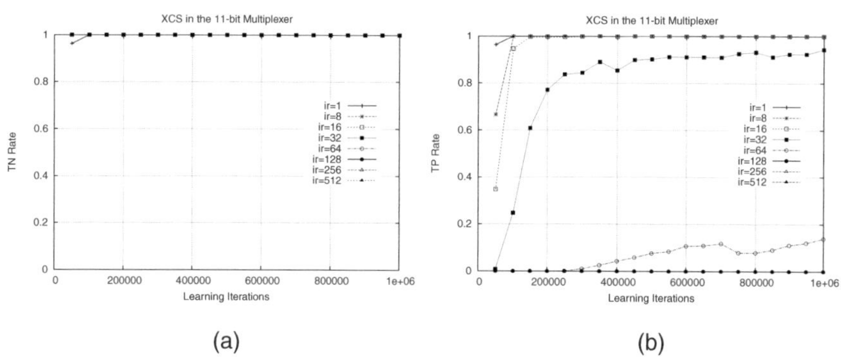

Fig. 1. TN rate (**a**) and TP rate (**b**) of XCS in the 11-bit multiplexer with imbalance ratios from $ir = 1$ to $ir = 512$

[1] For notation details, the reader is referred to [10, 32, 33].

Table 1. Most numerous rules evolved in a run of XCS with the 11-bit multiplexer for $ir = 128$. Cond. in the classifier's condition, A. the action it predicts, and p, ϵ., F and num are the prediction, error, fitness and numerosity of the classifier

Cond.	A.	p	ϵ	F	num
###########	0	1,000	$1.2 \cdot 10^{-4}$	0.98	385
###########	1	$1.2 \cdot 10^{-4}$	$7.4 \cdot 10^{-5}$	0.98	366

represented about 90% of the population, and for $ir = 256$, all the classifiers in the population were overgeneral. Moreover, for the highest ir, the error of overgeneral rules was lower than ϵ_0, and so, XCS considered these rules as accurate. For example, Table 1 shows that the error of the most overgeneral rules for $ir = 128$ is practically zero. In the next section, we theoretically analyze the effect of ir on the classifiers' error.

4 Modeling Parameter's Bounds

In this section, we theoretically relate the expected error of overgeneral classifiers with the imbalance ratio, and derive a bound on ir beyond which XCS will consider overgeneral classifiers as accurate. To derive the model, we assume that the imbalance ratio of the training dataset equals the imbalance ratio of the niches in the solution space (this assumption will be latter removed in Sect. 5). That is, we assume that instances of the minority class activate *starved niches*, and instances of the majority class trigger *nourished niches.* This assumption permits us to consider that there is a direct mapping between the imbalance ratio of the training set and the *small disjuncts* in the feature space.

4.1 Imbalance Bound

In balanced datasets, overgeneral classifiers will have a high error since they will cover, approximately, the same proportion of instances per class. Thus, the evolutionary pressures will discard them as long as more accurate classifiers exist in the population. However, as ir increases, these overgeneral rules receive less examples of the minority class, and so, they tend to have a lower error. At a given imbalance ratio, the error of these overgeneral rules will be less than ϵ_0; thus, they will be considered as accurate. We seek to derive the bound on the imbalance ratio to guarantee that overgeneral classifiers will be identified as inaccurate.

According to [8], the prediction p of a classifier can be approximated by:

$$p = P_c(cl) \cdot R_{max} + (1 - P_c(cl)) \cdot R_{min} \tag{2}$$

where $P_c(cl)$ is the probability that a classifier predicts the matching input correctly, R_{max} is the maximum reward, and R_{min} the minimum reward given by the environment. Then, the error of a classifier can be approximated by:

$$\epsilon = |p - R_{max}| \cdot P_c(cl) + |p - R_{min}| \cdot (1 - Pc(cl)) \tag{3}$$

For classification problems, R_{min} is usually 0, so that the prediction of a classifier can be estimated by: $p = P_c(cl) \cdot R_{max}$. Substituting p into (3), we get the following prediction error estimate:

$$\epsilon = 2R_{max} \cdot (P_c(cl) - P_c(cl)^2) \tag{4}$$

Now, let's relate $P_c(cl)$ with ir. In average, overgeneral classifiers will match ir examples of the majority class for each example of the minority class. Assuming that p is correctly estimated, a classifier would correctly predict the output for the ir instances of the majority class, and would give an erroneous prediction for the example of the minority class. Thus, $P_c(cl)$ can be approximated as:

$$P_c(cl) = \frac{ir}{1 + ir} \tag{5}$$

and its error estimate as:

$$\epsilon = 2 \cdot R_{max} \frac{ir}{(1 + ir)^2} \tag{6}$$

An overgeneral classifier will be considered inaccurate as long as:

$$\epsilon \geq \epsilon_0 \tag{7}$$

Using (6), we obtain that:

$$2 \cdot R_{max} \frac{ir}{(1 + ir)^2} \geq \epsilon_0 \tag{8}$$

which can be written as:

$$-ir^2\epsilon_0 + 2ir(R_{max} - \epsilon_0) - \epsilon_0 \geq 0 \tag{9}$$

This represents a parabola where ϵ takes values higher than ϵ_0 for ir ranging between ir_l and ir_u, where $ir_l < ir_u$. We are concerned about the maximum imbalance ratio up to which XCS would consider overgeneral classifiers as inaccurate; that is, ir_u. Solving (9), and assuming that $\epsilon_0 << R_{max}$, we obtain the following expression:

$$ir_u \approx \frac{2R_{max}}{\epsilon_0} \tag{10}$$

That is, the maximum imbalance ratio up to which XCS will be able to detect overgeneral classifiers depends proportionally on R_{max} and inversely proportionally on ϵ_0. Substituting $\epsilon_0 = 1$ and $R_{max} = 1{,}000$, the maximum imbalance ratio is: $ir_u \approx 2{,}000$. Nonetheless, our experiments at $ir = 128$ showed that the final populations evolved by XCS consisted mostly of overgeneral classifiers, indicating that XCS was not able to maintain accurate classifiers. In the following we analyze the potential causes of these deviations.

4.2 Theoretical and Experimental Bounds: Analysis of the Deviation

We investigate the potential causes of the deviation between the experiments and the theory. The theoretical bound shows that XCS should be able to learn up to an imbalance ratio of 2,000. However, the results obtained were far from this bound.

The theoretical bound is derived from (7), where we specify that the error of the classifier should be higher than ϵ_0. Thus, we assumed that the error of the classifiers is well estimated. Since XCS learns incrementally and updates parameters based on a windowed weighted average, the effect of some rewards may be forgotten if some examples come very infrequently. In the next section, we revise the method of error estimation and whether it can suffer from high imbalance ratios.

XCS's genetic algorithm is activated on a frequency basis, which may also be prone to high imbalance ratios. As ir increases, instances of the minority class are sampled more infrequently; so, starved niches (which are matched by these instances) are activated more infrequently with respect to nourished niches. This means that nourished niches tend to receive a higher number of genetic events, having more offspring which may overtake the population. This effect is uncorrelated with the parameter estimation. Even though parameters are well estimated, XCS's genetic algorithm may be responsible for the deviation between the experiments and the theoretical bound. Section 4.4 analyses the mechanism of occurrence-based reproduction and suggests ways to counterbalance the effects.

4.3 Learning Rate and Error Estimates

In XCS, classifier's parameters are adjusted using the standard *Widrow–Hoff delta rule* [31] with learning rate parameter β, where $0 < \beta \leq 1$ (see Sect. 2 for details). Thus, classifier's parameters are updated incrementally at learning rate β; they are expected to be stabilized to their theoretical average values after the classifier receives a certain number of updates. Nonetheless, in the experiments with high imbalance ratios, the parameters of overgeneral classifiers did not converge to their theoretical values. Note that, for $ir = 128$ (see Table 1), the population basically consisted of two overgeneral classifiers: (a) cl_1:`###########:0` with $P = 1{,}000$ and ϵ practically 0, and (b)

cl_2:###########:1 with both the error and the prediction really close to 0. The values of p and ϵ do not correspond to their theoretical ones, which can be approximated as follows (see (2) and (3)):

$$p_{cl_1} = 992.24 \qquad p_{cl_2} = 7.75 \tag{11}$$

$$\epsilon_{cl_1} = 15.38 \qquad \epsilon_{cl_2} = 15.38 \tag{12}$$

The difference between the theoretical estimates and the real values taken in the experiments may be due to the value of the learning rate β. In fact, β determines the amount of update in the parameter estimates. High values of β produce big corrections of the classifier's parameters. Usually, this allows for a faster convergence of the classifier's parameters to their real values. Low values of β would cause small corrections and so a slower convergence; for very small values, accurate offspring classifiers may loose against overgeneral parents at the beginning of the run, since their fitness increases slowly. In the reported experiments, we set $\beta = 0.2$, which is a typical value used for XCS [8, 32, 33]. In all these cases, XCS received a uniform distribution of samples, and so, classifiers of any class were updated with a similar frequency. In case of high imbalance ratios, overgeneral classifiers match one instance of the minority class every ir instances of the majority class. Consequently, high values of β may produce oscillations on the parameter estimates when the minority class instance is sampled. Small values of β may reduce the oscillations, but then the convergence time will be increased.

To check if the parameters of overgeneral classifiers oscillated, we monitored the error of the classifier ###########:0 along a single run of XCS for $\beta = 0.2$ and $\beta = 0.002$. The population was initialized with the two maximum overgeneral classifiers, and both covering and GA were switched off. The error of the classifier was sampled every 1,000 iterations. Figure 2 shows the distribution of the error for $\beta = 0.2$ (Fig. 2a and $\beta = 0.002$ (Fig. 2b)). For $\beta = 0.2$, the error of the classifier was almost zero most of the time, while the theoretical value is 7.75; moreover, there are some peaks in the distribution density, which indicate that the error momentarily changed when it received an example of the minority class, but quickly recovered the value of 0. Note that as $\epsilon \approx 0$, it satisfies that $\epsilon < \epsilon_0$, and so, the classifier is considered accurate most of the

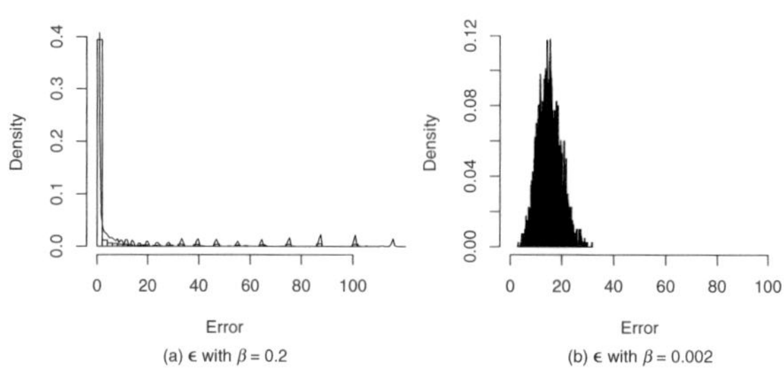

Fig. 2. Evolution of the parameter ϵ of the classifier ###########:0 in the 11-bit multiplexer at imbalance ratio $ir = 128$: (**a**) using $\beta = 0.2$; (**b**) using $\beta = 0.002$

time. Decreasing β to 0.002 smoothes the density curves and the distribution becomes closer to the theoretical value of the error. Although not shown for brevity, the same conclusions can be drawn for the classifier's prediction.

The experiments herein point out that the deviation caused by the online update of classifier parameters is significant for high imbalance ratios and high values of β. Thus, we could rewrite (7) but considering the deviation with respect to the theoretical bound as follows:

$$\epsilon \pm \sigma > \epsilon_0 \tag{13}$$

where σ is the maximum deviation in the parameters of overgeneral classifiers caused by the online update mechanism.

We decreased β to 0.002 to minimize the effect of the deviations and reran XCS with the 11-bit multiplexer. We found that the classifier parameters were better estimated, but the global TP rate was not improved as overgeneral classifiers persisted in the population for higher imbalance ratios. This indicates that there are more complexities affecting XCS for high class imbalances.

4.4 Occurrence-Based Reproduction

The imbalance ratio affects the proportion of reproductive opportunities that the different classifiers receive. As ir increases, *starved niches* are activated less frequently, and so, accurate classifiers that belong to these niches receive a minor number of genetic events. On the other hand, accurate classifiers that form *nourished niches* and overgeneral classifiers covering several input states get a higher number of genetic events. Thus, there is a disproportion, which increases with ir, on the number of genetic events that classifiers belonging to starved niches receive with respect to those of overgeneral classifiers and classifiers that belong to nourished niches. Since reproduction is niche-based, but deletion is population-based, an excessive disproportion may hinder starved niches from being evolved, and eventually, accurate classifiers contained in starved niches may be removed from the population. In this section, we theoretically model the genetic opportunities of these classifiers, and suggest a method to counterbalance this disproportion.

For this purpose, we focus on the reproduction opportunities that receive (a) accurate classifiers belonging to nourished niches, (b) accurate classifiers belonging to starved niches, and (c) the most overgeneral classifiers. As the selection in XCS is niche-based, we first compute the classifier's probability of belonging to an action set, which we denote as P_{occ}.

Instances of the minority class are sampled with probability $1/(1 + ir)$. As XCS chooses the class to explore randomly, the niches activated by these instances (referred as *starved niches*) are activated with the following probability:

$$P_{occ_{starved}} = \frac{1}{n \cdot m_s} \cdot \frac{1}{1 + ir} \tag{14}$$

where n is the number of classes, and m_s the number of starved niches. Similarly, recognizing that instances of the majority class are sampled with probability $ir/(1+ir)$, the niches activated by these instances (addressed as *nourished niches*) have the following probability of occurrence:

$$P_{occ_{nourished}} = \frac{1}{n \cdot m_n} \cdot \frac{ir}{1+ir} \tag{15}$$

where m_n is the number of nourished niches. Finally, the most overgeneral classifiers always participate in the match set, and the action set they advocate is randomly selected with probability 1/n:

$$P_{occ_{ovg}} = \frac{1}{n} \tag{16}$$

Once an action set is activated, the parameter update procedure is triggered, and the parameter values are adjusted according to the reward received. Thus, overgeneral classifiers and classifiers that belong to nourished niches would be updated more frequently, and so, they would have more reliable estimates. In the remainder of this analysis we consider that all classifier's parameters are accurate.

An action set receives a genetic event if the average time since the last application of the GA on this action set is greater than θ_{GA}. If the period of activation T_{occ} of a niche is higher than θ_{GA}, the classifiers that belong to that niche will receive a genetic event every time the action set is formed; thus, the period of application of the GA (T_{GA}) will be $T_{GA} = T_{occ}$. Otherwise, if $T_{occ} < \theta_{GA}$, the period of application of the GA will be: $T_{GA} \approx \theta_{GA}$.

The period of occurrence T_{occ} of the three types of classifiers is:

$$T_{occ_{starved}} = n \cdot m_s \cdot (1+ir) \tag{17}$$

$$T_{occ_{nourished}} = n \cdot m_n \cdot \frac{1+ir}{ir} \tag{18}$$

$$T_{occ_{ovg}} = n \tag{19}$$

Assuming that (a) ir is high (i.e., $ir/(ir+1) \rightarrow 1$), and (b) $m_n \cdot n \leq \theta_{GA}$, and not considering overlapping classifiers, we derive the period of GA application T_{GA} for the three types of rules:

$$T_{GA_{starved}} \approx n \cdot m_s \cdot (1+ir) \tag{20}$$

$$T_{GA_{nourished}} \approx \theta_{GA} \tag{21}$$

$$T_{GA_{ovg}} \approx \theta_{GA} \tag{22}$$

Note that the time between genetic events of starved niches increases linearly with the imbalance ratio and the number of classes. For the other classifiers, T_{GA} depends only on θ_{GA}. The relation between the number of genetic opportunities received by classifiers that belong to starved niches with respect to the other classifiers is:

$$\frac{T_{GA_{starved}}}{T_{GA_{nourished}}} \approx \frac{T_{GA_{starved}}}{T_{GA_{avg}}} \approx n \cdot m_s \cdot \frac{(1+ir)}{\theta_{GA}} \tag{23}$$

If $\theta_{GA} << ir$, overgeneral classifiers and classifiers that belong to nourished niches will receive an increasing amount of genetic opportunities with respect to classifiers belonging to starved niches, and so, they will have more offspring.

To counterbalance the number of genetic opportunities that the different niches receive, we require the following condition:

$$T_{GA_{nourished}} = T_{GA_{starved}} \tag{24}$$

Using (20) and (21), we obtain that:

$$\theta_{GA} \approx n \cdot m_s \cdot (1+ir) \tag{25}$$

In $O-$notation, and disregarding the effect of n and m_s:

$$\theta_{GA} = O(ir) \tag{26}$$

This indicates that θ_{GA} should be set according to the imbalance ratio ir to guarantee that nourished niches and starved niches will receive a similar proportion of genetic events. Also note that, under this condition, the classifiers belonging to nourished niches and overgeneral classifiers will receive more parameter's updates than the classifiers that belong to starved niches, and so, will have more accurate estimates.

4.5 Guidelines for Parameters Configuration

The above analysis provided insight into the role of some XCS's parameters and their influence on learning from imbalanced data. In the following, we derive some guidelines on how to set these parameters depending on the imbalance ratio.

First, ϵ_0 and R_{max} determine the maximum imbalance ratio up to which XCS will consider overgeneral classifiers as inaccurate classifiers (see (9)). Thus, these parameters set the threshold between negligible noise and imbalance. Regarding (7), if the error of an overgeneral classifier ($\epsilon > 0$) is smaller than ϵ_0, XCS will consider the classifier as accurate. Therefore, the few examples responsible for the error are considered as noise. Otherwise, if the error of an overgeneral classifier is higher than ϵ_0, XCS takes the classifier as inaccurate. So, the examples that make the classifier erroneous are considered as relevant examples and the classifier should not overgeneralize them.

XCS updates classifiers' parameters as a time-weighted average of their values. The learning rate parameter β adjusts the importance of the recent rewards in the parameter update mechanism. Low values of β mean that classifiers' parameters suffer small corrections every time they are updated. High values of β imply big corrections in classifier's parameters; so, parameter's estimates reflect the weighted average of few instances. In this case, we showed that the parameters of overgeneral classifiers can fluctuate, and so, overgeneral classifiers can be considered accurate during most of the learning time. Our suggestion to avoid this is to set β according to the activation frequency of the most starved niche (f_{min}) and the most nourished niche (f_{maj}), ensuring that the rewards provided when sampling instances of the minority class will be reflected in the parameters' values:

$$\beta = k_1 \cdot \frac{f_{min}}{f_{maj}} \tag{27}$$

where k_1 is an arbitrary constant. Under the initial assumptions of only having two types of niches, the starved and the nourished niches, the ratio of frequencies equals the inverse of the imbalance ratio: $f_{min}/f_{max} = 1/ir$. Thus:

$$\beta = \frac{k_1}{ir} \tag{28}$$

Finally, Sect. 4.4 argued that θ_{GA} should increase linearly with the imbalance ratio to ensure that nourished and starved niches received a similar number of genetic events. Generalizing, we write the following equation:

$$\theta_{GA} = k_2 \cdot ir \tag{29}$$

where k_2 is an arbitrary constant. For $k_2 = 1$, all niches will receive, approximately, the same number genetic events.

We ran the same experiments with the 11-bit multiplexer but setting XCS as indicated by the configuration guidelines. We only changed the parameters of the runs that failed: $ir = \{64, 128, 256, 512\}$. Specifically, we set $\theta_{GA} = \{200, 400, 800, 1{,}600\}$ and $\beta = \{0.04, 0.02, 0.01, 0.005\}$ for each imbalance ratio respectively. Figure 3 shows the results obtained. It can be observed that XCS solves the 11-bit multiplexer up to an imbalance ratio of $ir = 256$, which supposes a big improvement with respect to the initial experiments.

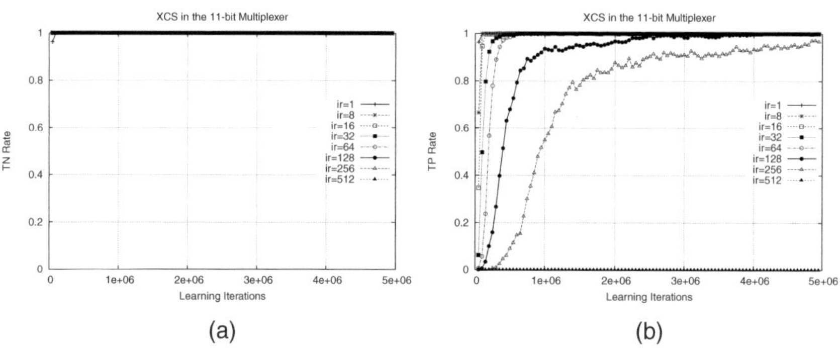

Fig. 3. (**a**) TN rate and (**b**) TP rate of XCS in the 11-bit multiplexer with imbalance ratios from $ir = 1$ to $ir = 512$. Parameters are configured according to the guidelines

The theoretical bound derived in Sect. 4.1 indicates that XCS might be able to solve the problem up to $ir = 2{,}000$. We got closer to this bound with appropriate parameter settings. However, we are still far from $ir = 2{,}000$. We hypothesize that the gap between the theoretical and the empirical maximum ir may be due to an insufficient number of classifiers of the minority class in the initial population. This would prevent XCS to settle the minority class niches and let them grow. As future work, we will analyze population sizing to guarantee the initial supply of classifiers belonging to starved niches at extremely high class imbalances.

5 Online Configuration of XCS to Handle Imbalanced Problems

The analysis in the last section was done assuming that there were only two types of niches in the solution space, the nourished niches and the starved niches, and that they were activated with a frequency directly proportional to the imbalance ratio. This was the case of the multiplexer problem. Nonetheless, in real-world problems, niches are unknown before running the system; consequently, niche frequencies cannot be estimated and may not be related to the imbalance ratio. In fact, the imbalance ratio reports about the proportion of examples per class, but does not provide any information about the distribution of the niches in the solution space. For example, even with a balanced dataset (i.e., $ir = 1$), there might be starved niches in the feature space.

Thus, we are concerned about the ratio between the frequencies of nourished niches and starved niches that lay closely in the feature space, rather than about the imbalance ratio of the training dataset. In this context, the guidelines proposed in Sect. 4.5 still hold, but now replacing the imbalance ratio of the training dataset ir by the *niche imbalance ratio* ir_n, defined as the ratio between the frequencies of the most nourished and the most starved niche. However, obtaining an accurate estimate of ir_n poses a big challenge to XCS, since the niches that XCS has to evolve and their frequencies are completely unknown in a real-world problem. Next, we present an approach, addressed as the *online adaptation algorithm*, that estimates ir_n from information gathered during XCS's learning; then, it substitutes this estimate in the formulas presented in the previous section to adapt XCS's parameters online.

5.1 Online Adaptation Algorithm

The online adaptation algorithm benefits from the potential information contained in the overgeneral classifiers to estimate the niche imbalance ratio ir_n of an unknown problem, and then use this estimate to tune β and θ_{GA}. Overgeneral classifiers are activated in different niches, which tend to be close in the solution space. From these overgeneral classifiers, we estimate ir_n with

Algorithm 5.1: Pseudocode for the *online adaptation algorithm*

Algorithm: OnlineAdaptation (cl <u>is</u> classifier)
if *cl* <u>is</u> *overgeneral* **then**
    $ir_n := \frac{exp_{maj}(cl)}{exp_{maj}+exp_{min}(cl)}$
    if ($ir_n < \frac{2R_{max}}{\epsilon_0} \wedge exp_{cl} > \theta_{ir} \wedge num_{cl} > \overline{num}_{[P]}$) **then**
        Adapt β (cl)
        Adapt θ_{GA} (cl)
    end
end

the relative imbalance ratio in the region of the feature space that they cover, by computing the ratio between the number of instances of each class that the overgeneral classifier matches.

Algorithm 5.1 shows the pseudocode for the online adaptation algorithm. After every parameter update, the algorithm is triggered for each classifier *cl* in the match set. The first condition of the algorithm checks if *cl* is overgeneral. In this case, ir_n is calculated as the number of instances of the majority class with respect to the number of instances covered by the classifier. Then, the algorithm checks if ir_n is smaller than the maximum imbalance ratio up to which XCS should be able to distinguish an overgeneral classifier (see (10)). If the condition is satisfied, it indicates that the overgeneral rule has a higher error than what is considered as negligible noise. In this case, if the classifier has sufficient experience and high numerosity, β and θ_{GA} are adapted following the guidelines derived in Sect. 4.5.

β is adjusted so that the real prediction value of the overgeneral classifier is close to the theoretical one. To do that, we consider the worst case: we suppose that the classifier receives one example of the minority class, and then, ir_n examples of the majority class. We compute the error value that the classifier will have after receiving these ir_n+1 instances with the correspondent β value using the following series:

$$p_{ir_{\beta_0}} = R_{max} \cdot (1-\beta) \tag{30}$$

$$\forall 1 < i \leq ir_{cl} : p_{ir_{\beta_i}} = \beta(R_{max} - p_{ir_{\beta_{i-1}}}) \tag{31}$$

If $p_{ir_{\beta_i}}$ is far from the theoretical value, we decrease β by a proportion $\zeta < 1$ and repeat the same process. Consequently, the algorithm guarantees that, even in the worst case, the estimate of the classifiers parameters will be close to their real values.

Finally, the algorithm uses ir_n to tune θ_{GA} by applying (29):

$$\theta_{GA} = k_2 \cdot ir_n \approx k_2 \cdot \frac{exp_{maj}}{exp_{min} + exp_{maj}} \tag{32}$$

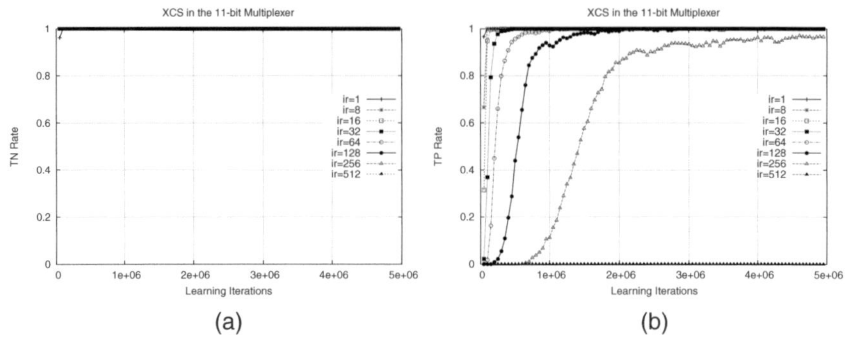

Fig. 4. (**a**) TN rate and (**b**) TP rate of XCS with online adaptation of parameters in the 11-bit multiplexer with imbalance ratios from $ir = 1$ to $ir = 512$

where k_2 fixes the minimum number of minority class examples that the starved niche has to match before going through a genetic event. This allows to counterbalance the genetic opportunities between starved and nourished niches. $k_2 = 1$ means that the classifiers belonging to starved niches are updated only once before receiving a genetic event. Higher values of k_2 allow better parameter estimates since these classifiers receive more updates between GA applications. In the experiments made in this section, we set $k_2 = 5$.

5.2 Results

Figure 4 shows the results obtained by XCS with online adaptation of parameters. The initial configuration reported in Sect. 3 was used. As β and θ_{GA} are adapted online, their initial value was set to $\theta_{GA} = 25$ and $\beta = 0.2$. The results are similar to those shown for XCS configured following the guidelines (see Fig. 3), where XCS could solve the 11-bit multiplexer up to $ir = 256$. With the online adaptation algorithm, the convergence is a little delayed since XCS needs some time to realize the existence of overgeneral classifiers, estimate ir_n, tune β and θ_{GA}, and let the evolutionary search remove overgeneral classifiers and discover accurate ones.

Let's note that this approach is essential in real-world problems, since there is not previous information about niche frequencies. In such a situation, the online adaptation algorithm introduces a significant improvement in imbalanced datasets.

6 LCS for Mining Imbalanced Datasets

XCS with online adaptation of parameters has demonstrated to be able to handle high amounts of class imbalance in artificially imbalanced problems. In this section, we investigate the capabilities of XCS for mining imbalanced data. Thus, we test XCS on a set of real-world problems with different imbalance ratios, and compare the system with other well-known machine learning techniques.

6.1 Methodology

We created a testbed consisting of 25 real-world two-class problems with different characteristics and imbalance ratios as follows. First, the following twelve problems were chosen: *balance-scale*, *bupa*, *glass*, *heart disease*, *pima indian diabetes*, *tao*, *thyroid disease*, *waveform*, *Wisconsin breast cancer database*, *Wisconsin diagnostic breast cancer*, *wine recognition data*, and *Wisconsin pronostic breast cancer*. All these problems where obtained from the UCI repository [7], except from *tao*, which was selected from a local repository [6]. For datasets with more than two classes, the discrimination of each pair of classes was considered as an individual problem. Thus, $\binom{n}{2}$ two-class problems were created from any problem with n classes (where $n > 2$), resulting in a testbed that consisted of 25 two-class real-world problems. Table 2 gathers the most relevant features of the problems.

The performance of XCS was compared to three of the most competent machine learning techniques: C4.5 [26], SMO [25] and IBk [1]. C4.5, derived from ID3, is one of the best representative decision trees which has been widely applied to tackle highly imbalanced problems. SMO is a fast method to train *support vector machines* [30] which has been able to handle very large training datasets [25]; in our experiments we used a linear kernel. IBk [1] is a nearest neighbor algorithm which decides that the output of a new input instance is the majority class of its k nearest neighbors; in the experiments, we set $k = 5$. All these machine learning methods were run using WEKA [34].

The metric of performance used in the comparison was the product of the TN rate and the TP rate, since this metric is not influenced by the imbalance ratio of the training dataset. To have good estimates of the TN rate and the TP rate, we ran the experiments on a ten-fold cross-validation [28]. After verifying that the results satisfied the condition of normality with the *Kolmogorov-Smirnov test* [29], the parametric statistical test of *repeated measures ANOVA* [27] was used to check if all the learning methods performed the same in average. Moreover, the performance of each pair of algorithms was compared using a *paired Student t-test* [28]. Next section shows the comparison among the four systems.

6.2 Results

Table 3 compares the performance of the four learners on the 25 datasets. The *repeated-measures ANOVA* did not permit to reject the null hypothesis that all the learners performed the same in average. This result is not surprising; in fact, the no-free-lunch theorem [35,36] justifies that, if no knowledge about the domain is used, no learning algorithm can systematically outperform the

Table 2. Description of the datasets properties. The columns describe the dataset identifier (Id.), the original name of the dataset (Dataset), the number of problem instances (#Ins.), the number of attributes (#At.), the proportion of minority class instances (%Min.), the proportion of majority class instances (%Maj.), and the imbalance ratio (ir)

Id.	Dataset	#Ins.	#At.	%Min.	%Maj.	ir
bald1	*balance-scale disc. 1*	625	4	7.84	92.16	11.76
bald2	*balance-scale disc. 2*	625	4	46.08	53.92	1.17
bald3	*balance-scale disc. 3*	625	4	46.08	53.92	1.17
bpa	*bupa*	345	6	42.03	57.97	1.38
glsd1	*glass disc. 1*	214	9	4.21	95.79	22.75
glsd2	*glass disc. 2*	214	9	6.07	93.93	15.47
glsd3	*glass disc. 3*	214	9	7.94	92.06	11.59
glsd4	*glass disc. 4*	214	9	13.55	86.45	6.38
glsd5	*glass disc. 5*	214	9	32.71	67.29	2.06
glsd6	*glass disc. 6*	214	9	35.51	64.49	1.82
h-s	*heart-disease*	270	13	44.44	55.56	1.25
pim	*pima-inidan*	768	8	34.90	65.10	1.87
tao	*tao-grid*	1,888	2	50.00	50.00	1.00
thyd1	*thyroid disc. 1*	215	5	13.95	86.05	6.17
thyd2	*thyroid disc. 2*	215	5	16.28	83.72	5.14
thyd3	*thyroid disc. 3*	215	5	30.23	69.77	2.31
wavd1	*waveform disc. 1*	5,000	40	33.06	66.94	2.02
wavd2	*waveform disc. 2*	5,000	40	33.84	66.16	1.96
wavd3	*waveform disc. 3*	5,000	40	33.10	66.90	2.02
wbcd	*Wis. breast cancer*	699	9	34.48	65.52	1.90
wdbc	*Wis. diag. breast cancer*	569	30	37.26	62.74	1.68
wined1	*wine disc. 1*	178	13	26.97	73.03	2.71
wined2	*wine disc. 2*	178	13	33.15	66.85	2.02
wined3	*wine disc. 3*	178	13	39.89	60.11	1.51
wpbc	*wine disc. 4*	198	33	23.74	76.26	3.21

others. However, we are interested in learners that are robust in average. For this purpose, we applied statistical pairwise comparisons (on a significance level of 0.99), which are shown as follows. The • and ∘ symbols indicate a significant degradation/improvement of the method with respect to another learner in the specific dataset. The last row of the table counts the number of times that a method has significantly degraded/improved the performance of another method.

Several observations can be drawn from the results. The overall degradation/improvement count shows that XCS is one of the most robust methods, specially for the most imbalanced datasets. Its performance is only degraded in eight occasions, the majority of which are concentrated in the problems: *bald2*, *bald3* and *tao*. *bald2* and *bald3* are two of the three datasets obtained from the discrimination of classes of the *balance-scale* problem. Both problems

Table 3. Comparison of the learning performance of C4.5, SMO, IBk, and XCS on the 25 real-world problems testbed. The • and ∘ symbols indicate that the correspondent learning algorithm performed significantly worst/better on a significance level of 0.99 (pairwise t-test) than another learning algorithm for the concrete problem. The row *Avg.* averages the performance of each method over all the 25 datasets, and the row *Score* counts the number of times that a method has performed worst-better than another for a specific problem

	C4.5	SMO	IBk	XCS
bald1	0,00	0,00	0,00	0,00
bald2	69,28 ••	83,98 ∘∘	81,16 ∘∘	71,22 ••
bald3	71,21 ••	85,69 ∘∘	82,11 ∘∘	70,07 ••
bpa	33,50 ∘	0,00 • • •	32,40 •∘	47,22 ∘∘
glsd1	79,60 ∘∘	0,00 ••	69,32 ∘	20,00 •
glsd2	33,95	15,00	24,13	59,40
glsd3	28,78	0,00	0,00	0,00
gls2c4	73,36	80,33	77,07	80,33
gls2c5	65,35 ∘	9,58 • • •	62,26 ∘	67,82 ∘
gls2c6	52,03 ∘	0,00 • • •	61,74 ∘	61,08 ∘
h-s	63,70	68,80	64,40	60,32
pim	44,96	48,36	46,91	46,06
tao	91,00 • ∘ ∘	70,57 • • •	94,25 ∘ ∘ ∘	82,90 • • ∘
thyd1	87,53	76,67	76,67	78,69
thyd2	93,12 ∘	54,17 •	77,90	82,50
thyd3	87,31 ∘	33,81 • • •	81,12 ∘	89,74 ∘
wavd1	67,80 • • •	78,65 ∘∘	72,28 • • ∘	80,43 ∘∘
wavd2	62,54 • • •	72,35 ∘∘	67,49 • • ∘	73,48 ∘∘
wavd3	68,61 • • •	79,61 ∘∘	74,14 • • ∘	81,01 ∘∘
wbcd	89,10	92,72	92,72	92,29
wdbc	88,83	94,27 ∘	93,47	90,30 •
wined1	85,58	98,46	94,98	99,23
wined2	91,83	97,51	97,50	99,17
wined3	87,64	97,14	87,94	93,43
wpbc	33,96 ∘	9,37 ••	28,98 ∘	20,99
Avg.	66,02	53,88	65,64	60,17
Score	14–10	20–11	7–16	8–12

have nearly the same proportion of instances per class ($ir = 1$). The reason why XCS is outperformed by SMO and IBk is not explainable, and cannot be caused by the imbalance ratio; thus, there may be other complexities affecting XCS's behavior. The *tao* problem is a completely balanced dataset ($ir = 1$) in which the boundary between classes is curved. In [4] it is shown that curved boundaries pose a challenge in XCS due to its hyperrectangular representation, which tends to concentrate high proportions of error.

On the other hand, XCS outperforms other methods in twelve occasions; from them, the datasets for which XCS outperforms more than one learner are: *bpa*, *wavd1*, *wavd2*, and *wavd3*. *bpa* is a quasi-balanced problem (ir = 1.38), and *wavd1*, *wavd2*, and *wavd3* are the three parwise discriminations of classes of the *waveform* problem, which present imbalance ratios of 2.02, 1.96, and 2.02 respectively. Besides, the *waveform* problem has the largest number of instances (5,000) and attributes (40) in the testbed. Although the difference between XCS and the other learners cannot be easily explained, these results indicate that XCS performs really competitively even when the size of the dataset increases. Let's also note that these results could be further explained by extracting the complexity of the training datasets. This approach was followed in [4, 5], evidencing a high correlation between some geometrical indicators of the training datasets and XCS's performance measured by the test error.

Let's now compare the learners in terms of imbalance robustness. To do that, we consider the problems with the highest ir; specifically, we analyze the performance on problems with $ir > 5$: *bald1*, *glsd1*, *glsd2*, *glsd3*, *glsd4*, *thyd1*, and *thyd2*. In all these problems, XCS performs really competitively. XCS is only outperformed in the problem *glsd1* by C4.5. In all the other problems, XCS performs equivalently to IBk and C4.5. SMO presents the worst performance. It has widely been shown in several works that C4.5 is able to deal with really imbalanced data [2, 19, 20]. Thus, the comparison indicates that XCS is robust (comparable to C4.5) for datasets with high disproportions of instances per class.

Finally, it is worth noting that there is not a direct mapping between the imbalance ratio in the learning dataset and the niche imbalance ratio, although both measures are related. Thus, even completely balanced datasets could cause *small niches* or *small disjuncts* to occur. As further work, we aim at designing metrics that, given a dataset, evaluate the presence of small disjuncts. This would allow to investigate the relationship between the classifiers' behavior and the presence of small disjuncts, and provide better understanding of the classifiers' performance on imbalanced datasets.

7 Summary and Conclusions

This work investigated the behavior of XCS on imbalanced classification problems. First, we empirically showed that XCS with a standard configuration can solve the multiplexer problem for moderate class imbalances ($ir \leq 32$). We identified that the number of overgeneral rules in the population tends to increase quickly with the imbalance ratio beyond a certain threshold of imbalance. For $ir > 64$, overgeneral rules represented near 100% of the population.

To provide further explanations on this tendency, we theoretically analyzed how the imbalance ratio affected the error of overgeneral classifiers, deriving a bound on the imbalance ratio under which XCS should be able to distinguish between accurate and overgeneral classifiers. As the theoretical bounds did not agree with the experimental observations, we analyzed the potential motifs of discordance. We detected that (a) the learning rate parameter β should be properly configured to ensure that overgeneral classifiers will have accurate estimates of their parameters, and (b) the GA (whose application rate is controlled by θ_{GA}) should be applied with a similar frequency to all niches to avoid that the offspring of classifiers that belong to nourished niches overtake the population. The analysis resulted in a set of recommendations on how to set β and θ_{GA} depending on the imbalance ratio, and results evidenced a significant improvement of XCS's behavior. We further argued the necessity of focusing on the niche imbalance ratio rather than in the imbalance ratio of the learning dataset to deal with real-world problems that may present small disjuncts. So, we proposed a method that estimates the niche imbalance ratio and automatically adjusts β and θ_{GA} from this estimate.

Finally, XCS was compared to C4.5, SMO and IBk on 25 real-world problems. The overall results showed that, although no learner performed statistically better than the others, XCS turned up to be really competitive to the other three machine learning techniques. The comparative analysis also denoted some differences in the performance of the learners that could not be easily explained by simply looking at the imbalance ratio. In fact, the imbalance ratio in the training dataset does not directly determines the presence of small disjuncts. Even completely balanced datasets can present small disjuncts depending on the distribution of instances around the feature space and the knowledge representation used by the learner. As further work, we propose to design metrics that evaluate the presence of small disjuncts, relating the performance of XCS with these indicators. Moreover, this information may be used in a corrective way, resampling the training dataset to diminish the presence of the small disjuncts.

Acknowledgements

The authors thank the support of *Enginyeria i Arquitectura La Salle*, Ramon Llull University, as well as the support of *Ministerio de Ciencia y Tecnología* under project TIN2005-08386-C05-04, and *Generalitat de Catalunya* under Grants 2005FI-00252 and 2005SGR-00302.

References

1. D.W. Aha, D.F. Kibler, and M.K. Albert. Instance-Based Learning Algorithms. *Machine Learning*, 6(1):37–66, 1991.
2. G. Batista, R.C. Prati, and M.C. Monrad. A Study of the Behavior of Several Methods for Balancing Machine Learning Training Data. *SIGKDD Explorations Newsletter*, 6(1):20–29, 2004.
3. E. Bernadó-Mansilla and J.M. Garrell. Accuracy-Based Learning Classifier Systems: Models, Analysis and Applications to Classification Tasks. *Evolutionary Computation*, 11(3):209–238, 2003.

4. E. Bernadó-Mansilla and T.K. Ho. Domain of Competence of XCS Classifier System in Complexity Measurement Space. *IEEE-TEC*, 9(1):1–23, 2005.
5. E. Bernadó-Mansilla, T.K. Ho, and A. Orriols-Puig. *Data Complexity and Evolutionary Learning: Classifierś Behavior and Domain of Competence*, pp. 115–134. Springer, Berlin Heidelberg New York, 2006.
6. E. Bernadó-Mansilla, X. Llorà, and J.M. Garrell. XCS and GALE: A Comparative Study of Two Learning Classifier Systems on Data Mining. In *Advances in Learning Classifier Systems*, volume 2321 of *LNAI*, pages 115–132. Springer, Berlin Heidelberg New York, 2002.
7. C.L Blake and C.J. Merz. *UCI Repository of Machine Learning Databases: http://www.ics.uc.edu/ mlearn/MLRepository.html.* University of California, 1998.
8. M.V. Butz, D.E. Goldberg, and T.K. Tharankunnel. Analysis and Improvement of Fitness Exploration in XCS: Bounding Models, Tournament Selection, and Bilateral Accuracy. *Evolutionary Computation*, 11(3):239–277, 2003.
9. M.V. Butz, K. Sastry, and D.E. Goldberg. Tournament Selection in XCS. In *Proceedings of the Genetic and Evolutionary Computation Conference (GECCO'2003)* , pages 1857–1869. Springer, Berlin Heidelberg New York, 2003.
10. M.V. Butz and S.W. Wilson. An Algorithmic Description of XCS. In P.L. Lanzi, W. Stolzmann, and S.W. Wilson, editors, *Advances in Learning Classifier Systems: Proceedings of the Third International Workshop*, volume 1996 of *Lecture Notes in Artificial Intelligence*, pages 253–272. Springer, Berlin Heidelberg New York, 2001.
11. P.K. Chan and S.J. Stolfo. Toward Scalable Learning with Non-Uniform Class and Cost Distributions: A Case Study in Credit Card Fraud Detection. In *Knowledge Discovery and Data Mining*, pages 164–168, 1998.
12. A. Van den Bosch, T. Weijters, and J. Van den Herik. When Small Disjuncts Abound, Try Lazy Learning: A Case Study. In *Proceedings Seventh BENELEARN Conference*, pages 109–118, 1997.
13. J.H. Holland. *Adaptation in Natural and Artificial Systems.* The University of Michigan Press, Michigan, 1975.
14. J.H. Holmes. Discovering Risk of Disease with a Learning Classifier System. In *Proceedings of 7th International Conference on Genetic Algorithms.* Morgan Kaufmann, San Mateo, 1997.
15. J.H. Holmes. Differential Negative Reinforcement Improves Classifier System Learning Rate in two-class Problems with Unequal Base Rates. In *Genetic Programming 1998: Proceedings of the Third Annual Conference*, pages 635–642. Morgan Kaufmann, San Mateo, 1998.
16. J.H. Holmes. *Rule Discovery in Epidemiologic Surveillance Data Using EpiXCS: An Evoluationary Computation Approach*, volume 358, pages 444–452. Springer, Berlin Heidelberg New York, 2005.
17. R.C. Holte, L.E. Acker, and B.W. Porter. Concept Learning and the Problem of Small Disjuncts. In *IJCAI'89*, pages 813–818, 1989.
18. N. Japkowicz. Concept Learning in the Presence of Between-Class and Within-Class Imbalances. In *Proceedings of the 14th Biennial Conference of the Canadian Society for Computational Studies of Intelligence*, pages 67–77. Springer, Berlin Heidelberg New York, 2001.
19. N. Japkowicz and S. Stephen. The Class Imbalance Problem: Significance and Strategies. In *IC-AI'00*, volume 1, pages 111–117, 2000.

20. N. Japkowicz and S. Stephen. The Class Imbalance Problem: A Systematic Study. *Intelligent Data Analisis*, 6(5):429–450, November 2002.
21. M. Kubat, R.C. Holte, and S. Matwin. Machine Learning for the Detection of Oil Spills in Satellite Radar Images. *Machine Learning*, 30(2–3):195–215, 1998.
22. A. Orriols-Puig and E. Bernadó-Mansilla. The Class Imbalance Problem in Learning Classifier Systems: A Preliminary Study. In *Genetic and Evolutionary Computation Conference (GECCO2005) workshop program*, pages 74–78. ACM Press, Washington D.C., USA, 25–29 June 2005.
23. A. Orriols-Puig and E. Bernadó-Mansilla. The Class Imbalance Problem in UCS Classifier System: Fitness Adaptation. In *Congress on Evolutionary Computation*, volume 1, pages 604–611. IEEE Press, Edinburgh, UK, 2–5 September 2005.
24. A. Orriols-Puig and E. Bernadó-Mansilla. Bounding XCS Parameters for Unbalanced Datasets. In *GECCO '06*, pages 1561–1568. ACM Press, Washington D.C., USA, 2006.
25. J. Platt. Fast Training of Support Vector Machines using Sequential Minimal Opt. In *Adv. in Kernel Methods - Support Vector Lear.* MIT, New York, 1998.
26. J.R. Quinlan. *C4.5: Programs for Machine Learning.* Morgan Kaufmann, San Mateo, 1995.
27. R.A. Fisher. *Statistical Methods and Scientific Inference*,2nd edition. Hafner Publishing Co, New York, 1959.
28. T.G. Dietterich. Approximate Statistical Tests for Comparing Supervised Classification Learning Algorithms. *Neural Computation*, 10(7):1895–1924, 1998.
29. T. Thode. *Testing for Normality.* Marcel Dekker, New York, 2001.
30. V. Vapnik. *The Nature of Statistical Learning Theory.* Springer, Berlin Heidelberg New York, 1995.
31. B. Widrow and M. Hoff. Adaptive Switching Circuits. In Insitute of Radio Engineers, editor, *IRE WESCON Convention Record*, volume 4, pages 96–104, 1960.
32. S.W. Wilson. Classifier Fitness Based on Accuracy. *Evolutionary Computation*, 3(2):149–175, 1995.
33. S.W. Wilson. Generalization in the XCS Classifier System. In *3rd Annual Conference on Genetic Programming*, pages 665–674. Morgan Kaufmann, San Mateo, 1998.
34. I.H Witten and E. Frank. *Data Mining: Practical Machine Learning Tools and Techniques*, 2nd edition. Morgan Kaufmann, San Francisco, 2005.
35. D.H. Wolpert. Stacked Generalization. *Neural Networks*, 5(2):241–259, 1992.
36. D.H. Wolpert. The Lack of A Priori Distinctions Between Learning Algorithms. *Neural Computation*, 8(7):1341–1390, 1996.

XCS for Fusing Multi-Spectral Data in Automatic Target Recognition

Avinash Gandhe[1], Ssu-Hsin Yu[1], Raman Mehra[1], and Robert E. Smith[2]

[1] Scientific Systems Company Inc., Woburn, MA, USA, syu@ssci.com, avinash@ssci.com
[2] RESystems Inc., Bokeelia, FL, USA; University College London, London, UK, robert.elliott.smith@gmail.com

Summary. We present our most recent efforts in applying XCS to automatic target recognition (ATR). We place particular emphasis on ATR as a series of linked problems, which include pre-processing of *multi-spectral* data, detection of objects (in this case, vehicles) in that data, and identification (classification) of those objects. Multi-spectral data contains visual imagery, and additional imagery from several infrared spectral bands. The performance of XCS, with robust features, notably exceeds that of a template-based classifier on the pre-processed multi-spectral data for vehicle identification. Promising preliminary results are also presented for vehicle detection. Future directions for this research are discussed in the conclusions.

1 Introduction

This chapter reviews our efforts in using XCS [16] in Automatic Target Recognition (ATR) [11, 12]. In our most recent efforts, we are developing a vehicle identification scheme that includes registration, detection and classification modules, all of which are important components of a higher-level, end-to-end ATR scheme.

In the remainder of this chapter, we describe our results to date. We first describe the multi-spectral data and hyperspectral data under investigation, and necessary pre-processing of that data. We then describe the XCS-based object detection and classification schemes we have developed. Finally, we present results based on ATR in multi-spectral images. XCS acts as a powerful selector of the pre-processed features for ATR detection and classification in this domain. XCS develops intuitive, highly-accurate, highly-general rules for the identification task. We present visualizations of vehicle detection (based on preliminary results). Final comments discuss future directions for our research.

Obtaining and pre-processing appropriate data is key to the ATR task. As we will see, raw, visual spectrum data is often inadequate, but multi-spectral data requires careful treatment. Such pre-processing develops the

A. Gandhe et al.: *XCS for Fusing Multi-Spectral Data in Automatic Target Recognition*, Studies in Computational Intelligence (SCI) **125**, 147–167 (2008)
www.springerlink.com

feature set which an algorithm like XCS can exploit. The following sections discuss the datasets we have employed, and the features we have developed via pre-processing.

2 Description of Datasets

We have focused on distinguishing different vehicle types (e.g., cars versus SUVs) from spectral data collected in an urban environment. In order to perform this task, the data sets we selected had to fulfill requirements in *spectral coverage* and in spatial *resolution*:

- *Spectral coverage* – Since vehicles can be of various colors, merely having visible spectrum (RGB) is not enough to distinguish between a vehicle and its background (e.g., we cannot distinguish between a green car and a patch of green grass). Thus we need broad spectral coverage to pick up the spectral signature of (metallic) vehicles.
- *Resolution* – In order to distinguish different vehicle types, we need to have resolution high enough to identify features of a vehicle (e.g., windshield, hood, truck, etc.). In practice, we need spatial resolution to be finer than 0.5 m.

2.1 Datasets Examined

We have examined a number of hyper-spectral and multi-spectral data sets to fulfill the requirements outlined above. Table 1 summarizes the data sets that we have obtained and examined.

A number of avenues were pursued to obtain data that satisfied our requirements. The first set of data investigated, the Washington DC Mall hyperspectral dataset, helped us to show good pixel classification and some target detection results based on spectral signatures of target and background classes. However, this data did not have sufficient resolution to support classification of vehicles by type. Of the four data sets, the RIT WASP data set has both the spectral coverage and the resolution that we need. This is the data set we will be using in our longer-term work. Meanwhile, the other data sets were still useful. For example, the Union Missouri data set was attractive due to the high resolution of the imagery and this data was used extensively to develop vehicle classification algorithms. Only the RIT WASP sensor data is described in some detail below.

2.2 RIT WASP Sensor Data

Since the RIT WASP data set is most relevant to our ongoing efforts, and illustrates the majority of our pre-processing efforts, we will focus on describing the pre-processing details of experiments conducted with the RIT WASP data.

Table 1. Summary of the data sets that we have examined

Data set	Spectral coverage	Resolution	Remarks
Washington DC Mall [6]	191 bands covering 400–2,400 nm	∼3 m	Produced by Purdue University. Used for initial experiments presented in August 2006
MassGIS [9]	Four bands (R,G,B, Near IR)	0.5 m	Produced by Massachusetts Office of Geographic and Environmental Information
RIT WASP [15]	Six bands (R,G,B, SWIR, MWIR, LWIR)	0.32 m for RGB, 2.0 m for the IR cameras	Produced by Rochester Institute of Technology
Union, Missouri	RBG and CIR (color-infrared)	0.15–0.3 m	Produced by SSCI. Covers the same area on different days, with different altitudes and various weather conditions

Table 2. Summary of camera capabilities

Camera	Spectral coverage	Frame size (pixels)	IFOV (mrad)	FOV (deg)	Frame rate (frames/s)
Visible (RGB)	400–900 nm	4,000 × 4,000	0.16	37.5	0.25
SWIR	900–1,700 nm	640 × 512	1.0	37.0	30 (max)
MWIR	3,000–5,000 nm	640 × 512	1.0	37.0	30 (max)
LWIR	8,000–9,200 nm	640 × 512	1.0	37.0	30 (max)

The Wildfire Airborne Sensor Program (WASP) is a multi-band mapping camera system developed by Rochester Institute of Technology (RIT) under a federal grant from NASA. As the name indicates, the sensor was originally developed to map wildfires from an airborne platform. The WASP camera system consists of four high performance frame cameras mounted on a common structure that pivots about a single axis to image a 4 mile swath from 10,000 ft. AGL. The camera suite provides simultaneous coverage of the electromagnetic spectrum from 400 to 9,200 nm. The capabilities are summarized in Table 2.

RIT has provided us with sample imagery from the WASP sensor that covers the RIT campus (see Fig. 1). Of particular note is the resolution of the cameras: 0.32 m for the visible RGB image and 2.0 m for the IR cameras. Since the images are of different resolutions, they need to be registered. The images contain a number of civilian vehicles. However, no ground truth is provided. Thus ground truth was generated by hand using the RGB image.

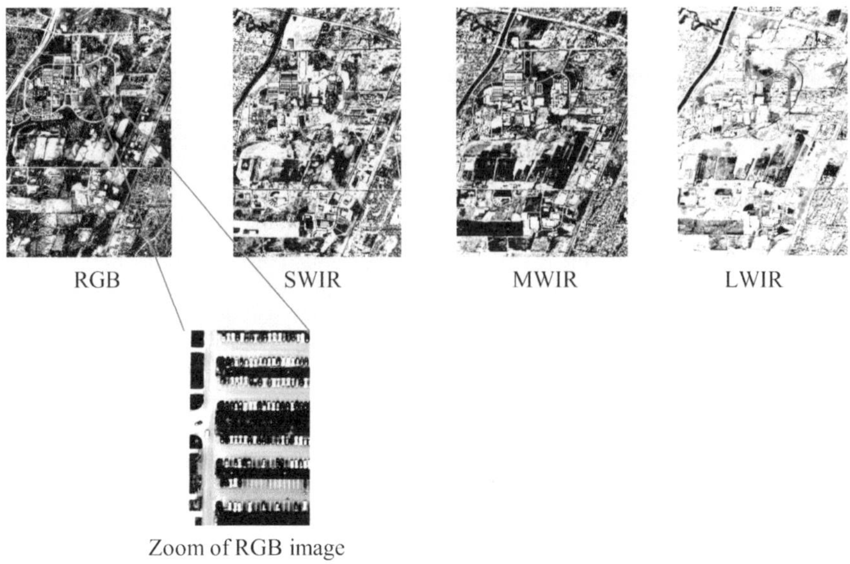

Fig. 1. RIT WASP sensor data (reproduced from [15])

3 Proposed Classification Scheme

The goal of the efforts presented here is to provide proof-of-concept of an end-to-end vehicle classification system using the XCS classifier as the underlying classification engine. In this section, we give an overview of the proposed system and then describe each algorithmic component in detail.

Overview

As shown in Fig. 2, our proposed vehicle classification scheme broadly comprises two stages: a *detection* stage and a *classification* stage. The goal of the *detection stage* is to isolate the objects of interest (vehicles) from the background. The *classification stage* will then classify the object into vehicle types. Depending on dataset used *registration* may be a necessary pre-processing stage.

The detection stage of the scheme will make use of both the spectral information provided via the IR bands of the dataset and the spatial resolution provided by visible wavelength (RGB) camera. While the IR bands may provide reliable information regarding the material at a certain pixel, at the resolution provided, the majority of vehicle pixels will comprise metal mixed with local background materials. Moreover, if two vehicles are close to each other the IR data may not be able to separate them. The RGB data on the other hand has high spatial resolution but high potential for false alarms since color information alone cannot distinguish cars from other similarly sized

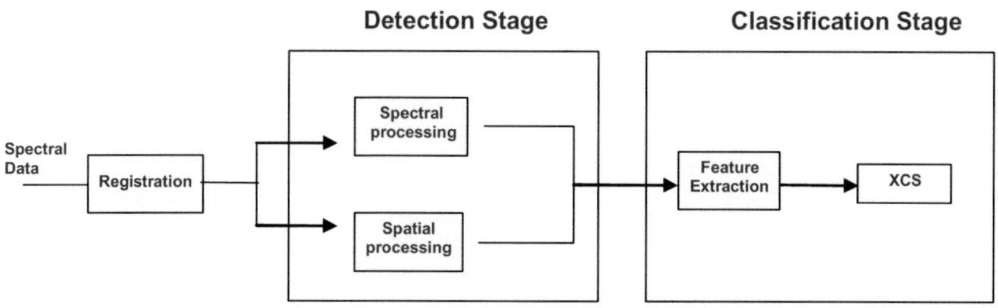

Fig. 2. Proposed Classification Scheme

objects. Therefore, the fusion of information from the RGB and IR bands is critical for detection. In our scheme, we will describe spatial processing that exploits the high resolution color information available and spectral processing that identifies the spectral signature of targets of interest.

Once targets of interest have been isolated, features must be extracted that will facilitate identification of the vehicle type. We describe a number of features used based on gradient, color and spatial information extracted from objects of interest. These features will then be used to train XCS to yield classification rules for each target class. The feature extraction and classification stage will make use solely of the RGB information since the IR images, with a native resolution of 2 m, simply does not provide enough information to classify targets.

We have developed the detection and classification stages of the proposed scheme independently. We will fuse the two stages in future work. In the following sections we will describe each algorithmic component in detail.

3.1 Registration of RIT WASP Data

Registration between the WASP images is necessary because the RGB and the IR cameras have different resolutions and cameras are not co-located.

Registration is carried out by computing a global affine transform. The underlying assumptions in such a computation are that (1) one image can be transformed into the other by some combination of rotation, translation and scaling, and (2) all points in the images lie in a common "ground-plane", i.e., they are at the same depth from the sensors. In order to register a pair of images, a number of *control points*, namely image points corresponding to the same physical, must be selected in each of the images. Since, there are only four images to be registered and the registration need be done only once, we hand-select the control points. An affine transform is computed between pairs of images using the control points by solving the following equation for the translation, rotation and scaling parameters:

$$\begin{bmatrix} \bar{x}_2 \\ \bar{y}_2 \end{bmatrix} = \begin{bmatrix} \alpha_{11} & \alpha_{12} \\ \alpha_{21} & \alpha_{22} \end{bmatrix} \begin{bmatrix} \bar{x}_1 \\ \bar{y}_1 \end{bmatrix} + \begin{bmatrix} t_x \\ t_y \end{bmatrix} \tag{1}$$

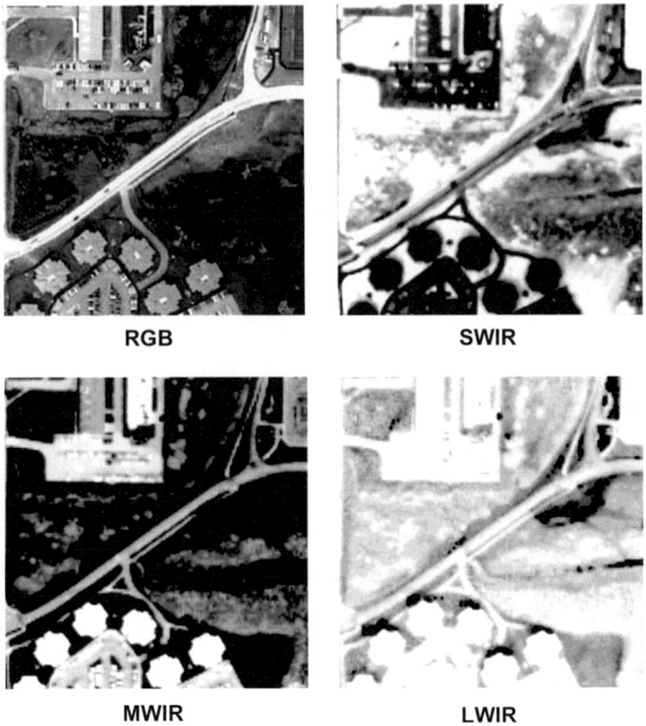

Fig. 3. Example of registration of sensor images

where, the pairs (x_1, y_1) and (x_2, y_2) are corresponding pixel locations in image 1 and image 2, α_{ij}, are the four parameters that encode rotation and scaling information and (t_x, t_y) are translation parameters. We conveniently use the MATLAB function *imtransform.m* to solve for these parameters. Once the affine transform is computed, one image can be transformed into the other. Since the RGB has the highest resolution, all the IR images are transformed to the footprint and ground spacing of the visible RGB image. Figure 3 shows a $1{,}000 \times 1{,}000$ footprint of the images resulting from the registration process, with all three IR images transformed to the RGB footprint shown.

3.2 The XCS Classifier

In our classification scheme of Fig. 2, we employ XCS to perform both spectral processing at the detection stage and vehicle identification at the classification stage. In this section we introduce our XCS classifier, discuss its implementation and describe a general XCS classification scheme that can be used for both pixel classification and vehicle identification.

4 XCS as a Learning Classifier System

A learning classifier system seeks to gain reinforcement from its environment based on an evolving set of condition/action rules called classifiers [7]. The environment associates with each classifier in the system a *reward* that contributes toward the classifiers *strength* each time it is encountered during the learning process. The goal of the system is to accumulate as much reward as possible by filling its population set with classifiers of high strength. The strength acts as a prediction of the reward the system receives each time that classifier is used. For this reason the strength is also termed a *reward prediction*. Rule discovery in classifier systems is achieved through the use of a Genetic Algorithm (GA).

The XCS was introduced by Wilson [2, 16, 17] as an enhancement to the traditional Learning Classifier Systems. Although details are not provided here, the traditional LCS uses the classifier strength as both the predictor of future payoff (prediction reward) and as the classifiers fitness in the genetic algorithm component of the classifier system. In the XCS, the classifier fitness is computed as a function of accuracy with which the classifier predicts the reward rather than strength parameter. The effect of using such an approach is that *consistently rewarded* classifiers are maintained in the GA population, regardless of their level of reward.

The XCS variant of LCS takes into account these characteristics and develops classifier rules that reflect both maximum accuracy and maximum generality. Consider the operational cycle of the XCS-based system we have constructed here, as it is being trained.

Initialize:

Set up a set X_0 of m points from the space of generalized observation-action pairs.

Iterate from t = 1, to Convergence:

1. Get the observation, z_t from the environment.
2. Set M_t (the match set) to the set of individuals in X_{t-1} that match observation z.
3. for $j = 1$ to m
   a. Determine the action(s) to be taken using the *reward-prediction* value associated with the individuals in M_t.
   b. Set A_t (the action set) to the set of individuals in M_t that have this action.
   c. Update the *reward-prediction* value for all elements $\chi_{t-1} \in A_{t-1}$.
   d. Update the *accuracy* (of *reward-prediction*) value for A_{t-1}.
   e. Increment a counter T in every individual in A_{t-1}.
   f. If the mean of T in A_{t-1} is greater than some threshold
      i. Add all elements of A_{t-1} to $X_{\mathrm{t}}^{\mathrm{AUX}}$.
      ii. Resample: for $j = 1$ to $||\mathrm{A}_{t-1}||$.

1. Draw $\chi_t^{[J]}$ from X_t^{AUX} with probability proportional to *accuracy*.

iii. Set $X_t = \Phi$

iv. Perform recombination and mutation over the elements of X_t^{AUX} to form new individuals.

v. Add the elements of X_t^{AUX} and some elements of X_{t-1} to X_t.

To see how this process balances a search for generalization and accuracy, note that the fundamental pressure is towards more accurate rules (in step (f) (ii) (1)), but also note that more general rules, which match more observations, will more often appear in the match set M (and thus the action set A) and then receive more reproductive opportunities in step f.

In our work we have used the freely available XCS software from the Illinois Genetic Algorithms Laboratory [4] and tailored it to our particular application. Examples using the XCS to generate classification rules are shown in the following sections.

5 Feature Representation

The formulation of rules in the XCS based on using the following representation:

IF <feature$_1$ generalized condition> AND <feature$_2$ generalized condition>, ...

<feature$_N$ generalized condition>

THEN <target class>

Similarly, specific data samples must be used to train the XCS by representing them as <condition>/<action> pairs of the form:

<feature$_1$ value>, <feature$_2$ value>, <feature$_N$ value>/<known target type>

In general, in order to train the classifier, each of the features used is represented by an n-bit string. We illustrate our feature representation scheme by using the example of training a classifier to distinguish two classes, grass and trees, based on their returns in two hyperspectral bands. The response in each band is considered a feature and a scatter plot of the samples from each class in the 2-D feature space are shown in Fig. 4. In order to create an n-bit representation of each feature, the feature space in each dimension must be divided into 2^n bins. The binning for a two-bit representation is shown in Fig. 4. The bins may be manually defined or created automatically using the mean and variance of all the training samples for a particular feature.

Once the feature space is divided into bins, each data sample can be transformed to the required <condition><action> representation. For instance, in the example shown in Fig. 4, if grass is class 1, then the highlighted data sample can be represented as <**01 10**><**1**> since it falls into the second bin

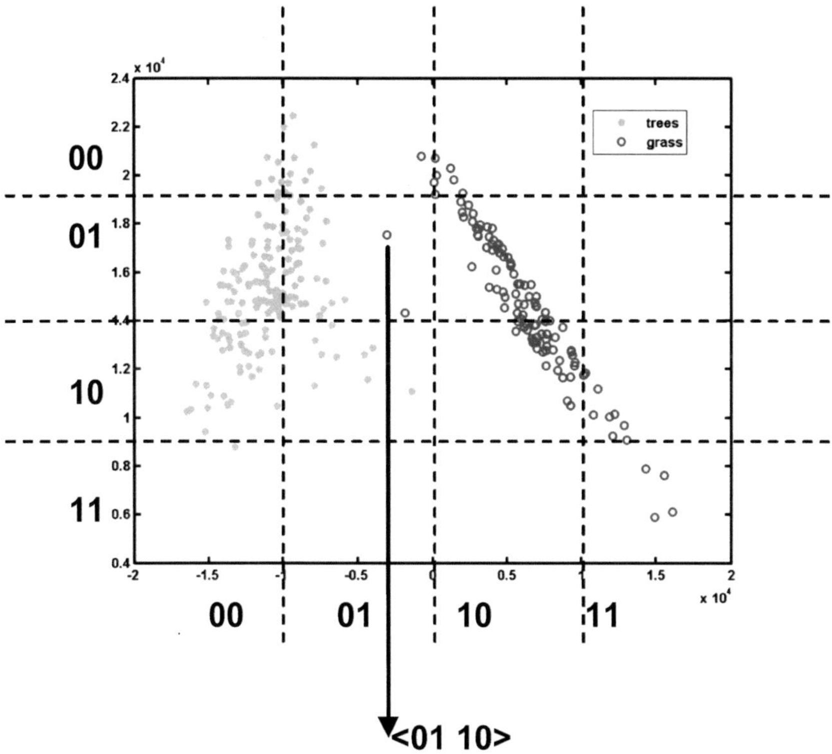

Fig. 4. Example of feature representation for a 2-D feature space

(**01**) of the first feature and the third bin (**10**) of the second feature. All data samples are similarly encoded.

6 XCS Based Classification Algorithm

An overview of a typical XCS-based classification algorithm is shown in Fig. 5. The training data, represented as <condition><action> pairs using the representation scheme in Sect. 5, are used by the XCS to generate *classification rules*. The classification rules are <condition><action> pairs with associated *fitness* (an inverse function of accuracy), f, and prediction reward, p. The XCS may generate hundreds of rules of varying fitness and reward, many of which may be conflicting. Due to the use of accuracy as a measure of fitness in the GA, XCS also generates *complementary* rules. These rules correspond to logical complements that indicate the absence of a certain class. The rules follow the rule representation described in Sect. 5 but have low prediction reward. The interpretation of such rules is: If <condition> then *not* <action>.

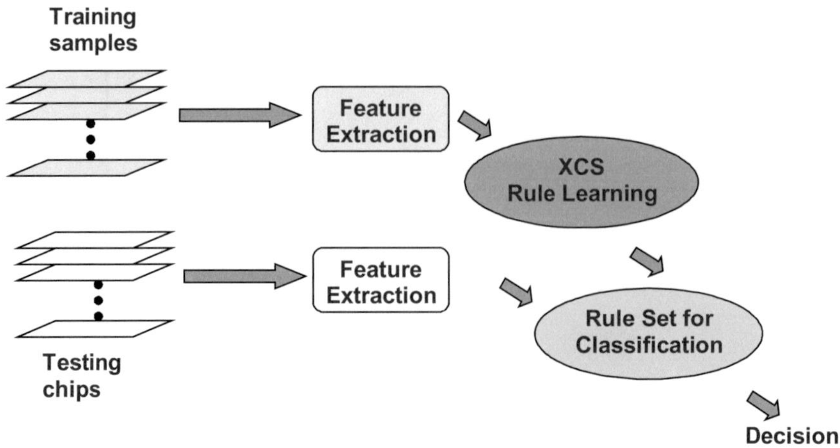

Fig. 5. XCS based classification algorithm

In the testing phase of the algorithm these rules are used to match test data (as represented in the previous section) and classify them accordingly into one of the candidate targets. Since a number of rules may match a particular test sample and some of these may provide conflicting <action> recommendations, the fitness weighted prediction reward is used to determine the classification.

For instance, when a test sample matches a set of N_i rules whose action is class i, the measure of similarity to that class is computed as follows:

$$L_i = \frac{\sum_{n=1}^{N_i} p_n f_n}{\sum_{n=1}^{N_i} f_n} \tag{2}$$

The class with the largest likelihood is chosen as the decision class and confidence of the decision in choosing class c is given by

$$c = \frac{L_c}{\sum_i L_i} \tag{3}$$

Note that the intrinsic availability of this confidence information is an important facet of what XCS offers to ATR.

7 Detection

As shown in Fig. 2 our detection scheme will fuse both the spectral and spatial information to extract objects of interest. Our approach consists of finding detection "masks" from both the spatial and spectral processing, and then fusing these masks to isolate objects of interest. Although the results presented in this section are only preliminary, they are promising, and we believe the methodology used is a well-suited part of the end-to-end data processing of the ATR solution suggested here. Details are provided below:

7.1 Spatial Processing

Our approach to spatial processing for object detection is based on simple color processing and spatial filtering. The steps in the procedure are listed below and illustrated by means of the example in Fig. 6.

1. Convert the RGB image to the YIQ color space. The YIQ color space separates the luminance information (Y-component) from the chrominance information (I-Q components) in an image.
2. Discard the luminance information. We are only interested in the basic color distribution of pixels in the target pixels and therefore attempt to remove the effects of reflections, shadows, glare, etc. by removing the Y-component of the YIQ image.
3. Apply an entropy filter. The entropy filter computes the local entropy around each pixel in the image. The local entropy gives a measure of the variance in the target which can be useful in distinguishing it from background.

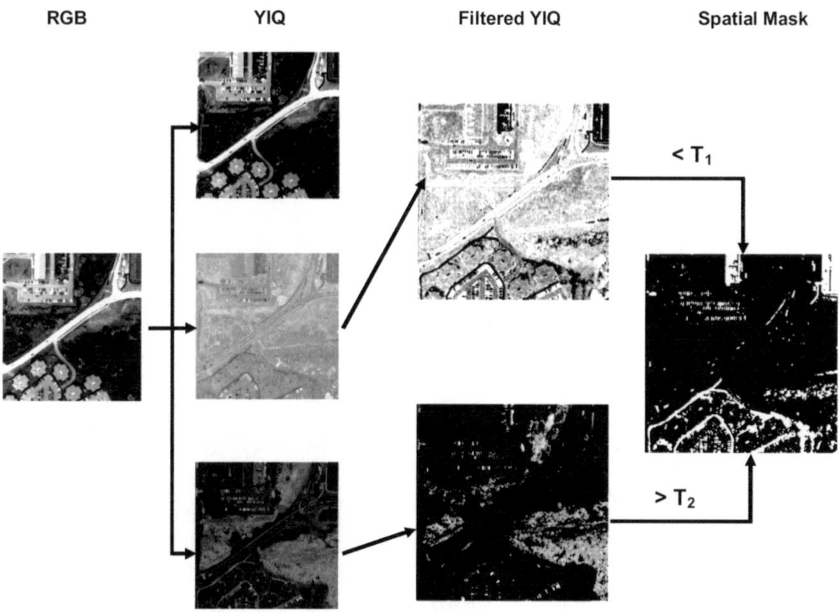

Fig. 6. Procedure to generate a spatial detection mask

4. Threshold the filtered I and Q components. It was found empirically that maintaining all information below a certain threshold in the filtered I-component and all information above a certain threshold in the filtered Q component resulted in retention of almost all the vehicles in the images.

The spatial detection mask resulting from the proposed scheme is shown at the end of the processing chain in Fig. 6. While a large amount of background clutter is retained, most of the background has been segmented out and all the targets of interest have been maintained. Since this is only one component of the detection algorithm the large amount clutter retained is not a significant issue as long as no targets are lost.

7.2 Spectral Processing

Our approach to using spectral information is based on pixel classification. The goal is to label each pixel in the scene as either a target or background pixel. In order to do so, we use the XCS classifier to generate classification rules for pixels. As in the case of spatial processing, we illustrate our scheme by showing results on a single 1,000 $\times$ 1,000 pixel region of the scene. The scene contains various background types and a number of vehicles. We begin by ground-truthing a number of pixels in the scene which are subsequently used to train the classifier. Figure 7 shows the areas of the scene used to train the classifier. The accompanying table also shows the number of samples from each class used.

In order to illustrate the information we expect to leverage to form a mask based on spatial processing, we show in Fig. 8 a scatter plot of the various samples in the LWIR and MWIR spectral bands. The figure shows that based on IR information alone a vehicle is well-separated from various

Class	Samples
Roof	2500
Road	2500
Path	480
Grass	2500
Trees	2500
Parking lot	2500
Vehicles	1297

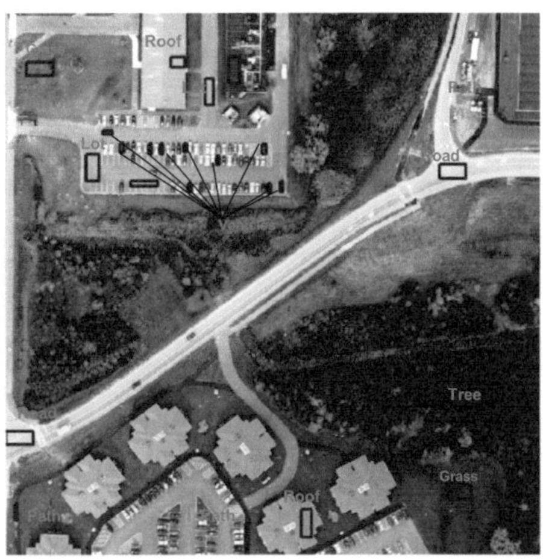

Fig. 7. Samples used for training in pixel classification approach

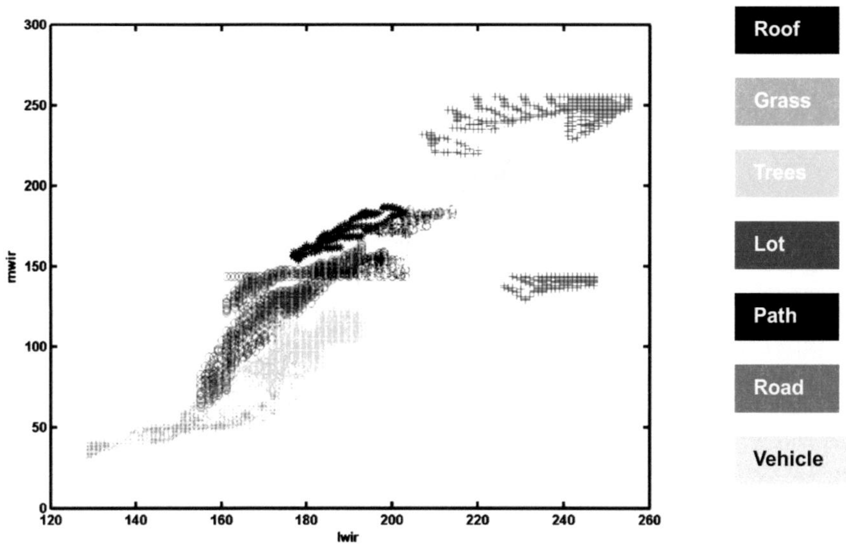

Fig. 8. Scatter plot of samples from various classes in LWIR-MWIR bands

background classes. There is a small overlap between the "parking-lot" class and the "vehicle" but this is to be expected considering the low resolution of the IR images which results in spectral mixing of these two classes in a large number of pixels that we have chosen to represent the vehicle class.

We train an XCS based on five attributes, each encoded with six bits. These attributes (for each training pixel) are (1) the I-component of the YIQ image, (2) the Q-component of the YIQ image, (3) the LWIR response, (4) the MWIR response, and (5) the SWIR response. While we have extracted samples from various background types, we treat all these samples as a single class (class 0) and the vehicle is the second class (class 1). Table 3 shows the significant parameters in the XCS and the values used in our application.

Some of the commonly used rules generated by the XCS are shown in Table 4 below.

The rules generated by XCS can be interpreted fairly easily. The first rule indicates that if the Q-component is large and the MWIR response is large, the pixel is from a vehicle. The second rule indicates that if the MWIR response is small, the pixel is from background. The fourth rule is an example of the complementary rules generated by XCS and indicates that if the I component is very large, the Q-component is large and the MWIR response is large, then the pixel is *not* from background.

The rules generated by XCS are applied to the entire $1{,}000 \times 1{,}000$ image and the resulting spectral detection mask is shown in Fig. 9. The foreground in the mask is made up of pixels classified as "vehicles" by XCS. As with the spatial detection mask described earlier a number of clutter objects are detected, however the targets of interested are retained.

Table 3. Parameter settings in XCS

Parameter	Setting
Max number of steps	50,000
Max population size	8,000
ε (tolerated error in prediction error)	0.2
Payment range	1,000
α (parameter to decrease accuracy value if prediction error is high)	0.1
β (learning rate parameter)	0.2
γ (parameter that discounts the max of prediction array)	0.71
Fitness reduction in offspring	On
Tournament size	0.4
Crossover probability	0.8
Mutation probability	0.04
GA subsumption	On
Action set subsumption	Off

Table 4. Rules generated by XCS for spectral classification

Condition					Action	Reward
######	1#####	######	1#####	######	1	1,000
######	######	######	0#####	##01##	0	1,000
1#####	###1##	######	1#####	######	1	1,000
11####	1#####	######	1#####	######	0	0
I	Q	LWIR	MWIR	SWIR		

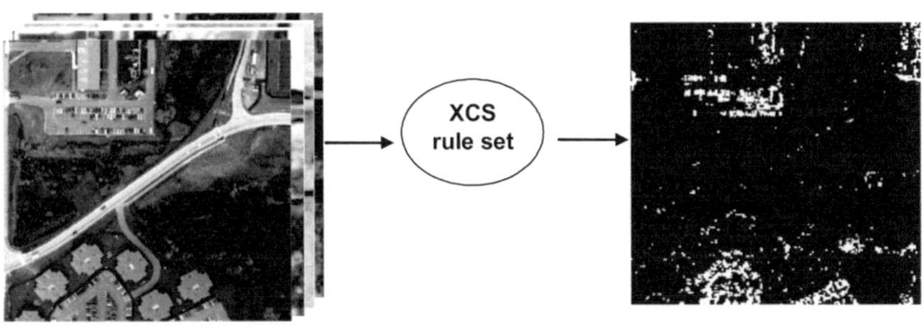

Fig. 9. Spectral detection mask generated from XCS rules

The spectral and spatial detection masks are combined as shown in Fig. 10. The masks are combined such that the resulting image only contains detections when the pixel is classified as a "vehicle" in both masks. For the binary masks, this is equivalent to taking the product of the two masks at each pixel. The resulting mask is then filtered to reject very large or very small objects and then applied to the original image to find objects of interest. The figure also shows close-ups of the original data and detection results in one of the

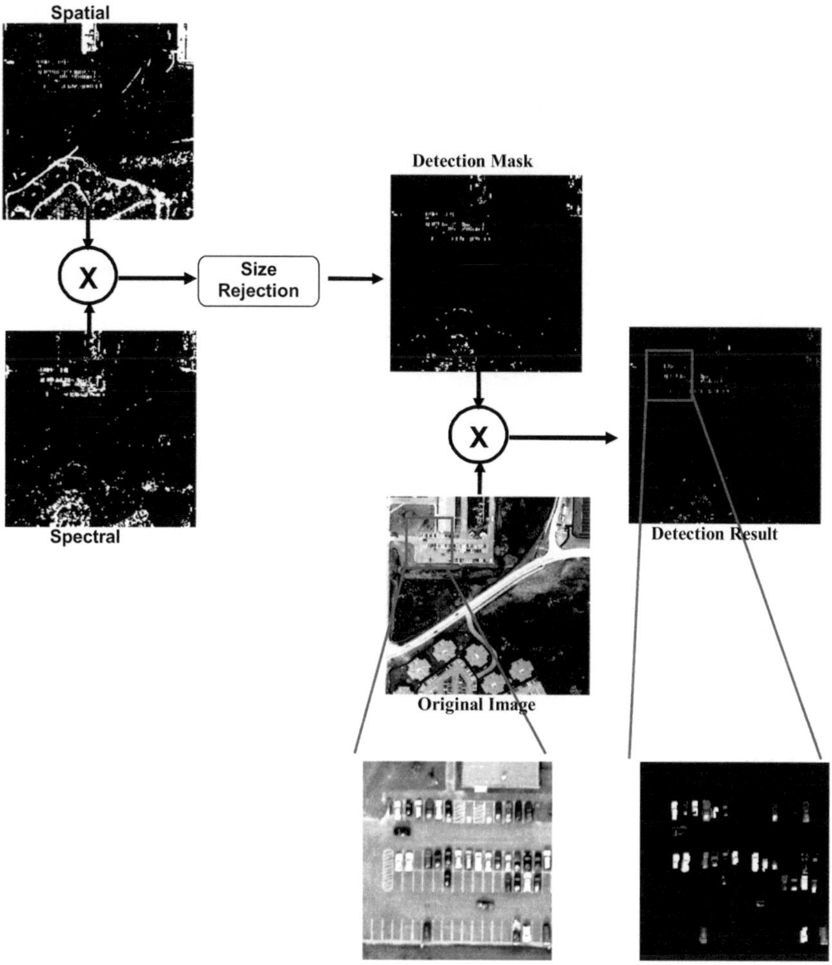

Fig. 10. Fusion of spectral and spatial masks

parking lots in the scene. The results show proof-of-concept of our approach to fusing spectral and spatial information for detection.

7.3 Classification

Once targets of interest have been detected they must be classified according to vehicle class. In the work presented here, we broadly define three vehicle classes: cars, SUVs and pick-up trucks. The resolution of the data and our inability to ground truth many of the vehicles restricts the number of vehicle classes and the diversity of the vehicles within each class. For instance, it is difficult to tell the difference between SUVs and minivans and therefore these are treated as one class.

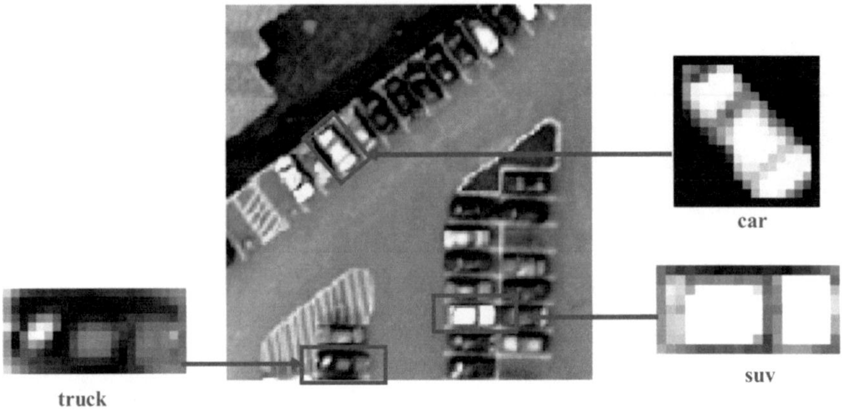

Fig. 11. Examples of vehicle types in RIT WASP data

Feature Extraction

Figure 11 shows examples of different vehicle types commonly encountered in the RIT WASP data. We have defined three classes: cars, SUVs and trucks and show examples of these in the figure. The image shows an (approximately) $45\,\mathrm{m} \times 45\,\mathrm{m}$ scene; note the difficulty in visually identifying the vehicles types due to low resolution. Also note that vehicles may appear in many different colors and at arbitrary angles. Features commonly used in classification of such objects are based on spatial properties such as shape and size. However, since we must base our classification on aerial views of the targets, given the wide range of sizes and shapes of civilian vehicles merely considering spatial features may not be adequate. One may notice that the distinguishing feature between different vehicle types is typically the number and spatial distribution of body components and sloped windshields. Cars typically have two sloped windshields while SUVs typically have one (depending on observation angle, a second, smaller one may be visible at the rear of the vehicle). Pick-up trucks commonly have a large component in the rear of the vehicle (the "bed", which often has a "bed liner") which is of different color from the front of the vehicle (the "cabin"). The transitions between windshields and body components affect the number and position of gradients in the image. Also, since vehicles are typically symmetric along width, we need only extract gradient features along the lengths of the vehicle.

In order to extract the features in the object of interest chips we must first detect gradients in the image. As mentioned earlier, since vehicles are largely symmetric along the width direction we are only interested in the gradients along the length. The first step is to detect the major axis of the object and rotate the chip such that its major axis is horizontal. We accomplish this using the MATLAB commands to detect region properties (*regionprops.m*) and rotate images (*imrotate.m*). An example is shown in Fig. 12.

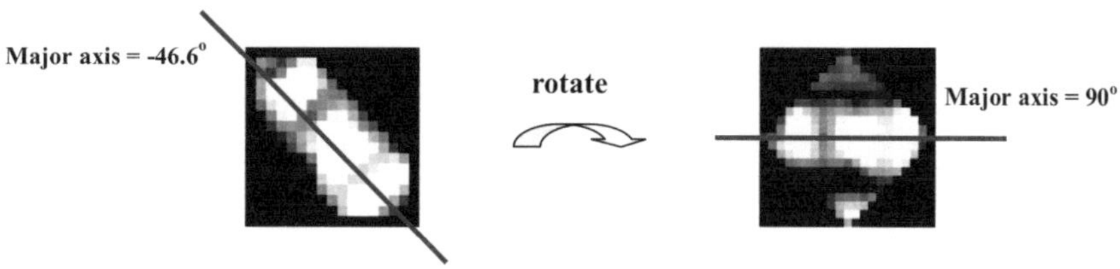

Fig. 12. Rotation of object mask

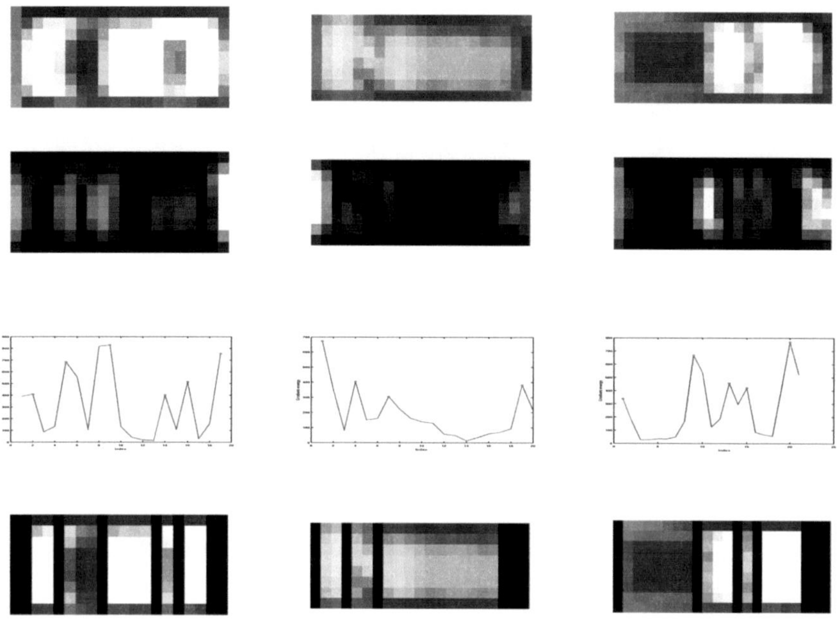

Fig. 13. Gradient detection and projections onto the major axis of the vehicle

We are interested in detecting only vertical edges in the rotated object mask. We initially experimented with the Orientation Energy approach described in [6] to find oriented gradients. However, we found that at the resolutions provided using a simple differencing along the horizontal direction (i.e., along each row of the rotated mask) provided best results. The results are shown in Fig. 13. As shown in the figure, the detected edges are also projected onto the major axis to form a gradient energy plot. The peaks of this projection (represented by the red '*'s) typically represent the transitions between vehicle *components* (roof, hood, bed, windshields, etc.). The vehicle components can then be extracted as the image area between peaks of the gradient projection. A representative color (RGB) is assigned to each component by computing the median of the colors at each pixel in the component.

The oriented gradients and vehicle components, once formed, must be transformed to a compact feature set that represents the spatial distribution

Table 5. Defined features used as attributes to train a classifier

Gradient Ratio: (GradRatio)	Ratio of gradients on each side of the center of the vehicle. This is computed by finding the center of gradient projection plot and summing the gradients on each side of the center. The gradient ratio is then the ratio of higher number to lower number.
Size of the largest component: (SizeLarge)	The length (along the major axis) of the largest component of the vehicle.
Color difference at ends: (ColDiff)	The mean squared difference between the colors of the two end components of the vehicle.
Size ratio: (SizeRatio)	The ratio of the size of the larger of the end components to the size of the smaller end component.

Table 6. Sample XCS rules for vehicle classification

Condition				Action	Reward
0###	####	0###	0###	0	864.6
11##	####	0###	####	1	1,000.0
####	####	1###	####	2	950.0
####	####	1###	####	1	0.0
GradRatio	*SizeLarge*	*ColDiff*	*SizeRatio*		

of the edges, colors and body components. Table 5 shows the features that were used as attributes to train classifiers:

Vehicle Classification

The features described above were used to train an XCS classifier. Three action classes were defined: car (class 0), SUV (class 1) and truck (class 2). A set of 40 car chips, 30 SUV chips and 20 truck chips were repeated with perturbation (Gaussian noise was added) to form a training set with 500 samples in each class. Each feature was binary encoded using four bits and the maximum reward provided is 1,000. The XCS generated a number of rules for classification of the vehicles. Examples of high-fitness, frequently-used rules for each class are shown in Table 6.

The rules can be interpreted as follows:

- The first rule shows that small values of *GradRatio, ColDiff* and *SizeRatio* indicate the presence of a car. Cars usually have two visible windshields and are fairly symmetric along the major axis, hence the small *GradRatio.* Also, the hood and trunk are always the same color and are of comparable size, hence the small values of *ColDiff* and *SizeRatio.*

- The second rule shows that a large *GradRatio* and small *ColDiff* indicates the presence of an SUV. This is true, due to the fact that only one windshield is commonly visible in an SUV and therefore there are more gradients on one side of the vehicle than the other. However, this feature alone is not sufficient since many trucks may have similar gradient ratios. Therefore, the small *ColDiff* resolves the trucks and SUVs since the roof of SUVs and the hood are the same color while the hood and bed of the pick up trucks are commonly of different color.
- The third rule shows that high values of *ColDiff* indicate the presence of a truck. One of the defining features of a truck in our data is the large *ColDiff* since the hood of the truck and the bed have different colors.
- The fourth rule shown is an example of a complementary rule which states that if the vehicle has high *Coldiff* it is definitely *not* an SUV. This rule is also consistent with the physical properties of an SUV.

The rules generated by the XCS were used to classify vehicle samples in a testing set. The testing set consisted of 40 car samples, 30 SUV samples and 20 truck samples. The results of classification (in %) of both the training and testing sets are shown in confusion matrices of Table 7.

These results are compared to the results shown in Table 8 obtained by using a a classifier based on template matching that uses known target chips to construct templates of each target. The templates are based on the gradient maps rather than the raw image chips. Test chips are then assigned to a target class by comparing the Mahalanobis distances [12] to each of the target templates. The results with both classifiers are comparable on the training set. However, on the testing sets, the XCS classifier outperforms the ML classifier by a significant margin, especially in the classification of SUVs and trucks. Based on a preliminary analysis of the results, we believe that the ML classifier over-trains and therefore shows lack of robustness. The XCS on the other

Table 7. Results for classifying vehicles using XCS

	Training			Testing		
	Car (%)	SUV (%)	Truck (%)	Car (%)	SUV (%)	Truck (%)
Car	97.5	0.0	2.5	97.5	0.0	2.5
SUV	0.0	100	0.0	3.3	93.3	3.3
Truck	5.0	0.0	95.0	10.0	0.0	90.0

Table 8. Results for classifying vehicles using template-matching

	Training			Testing		
	Car (%)	SUV (%)	Truck (%)	Car (%)	SUV (%)	Truck (%)
Car	100	0.0	0.0	87.5	5.0	7.5
SUV	3.3	96.7	0.0	23.3	70.0	6.7
Truck	5.0	0.0	95.0	30.0	5.0	65.0

hand, due to its inherent generalization mechanism and training procedure, generalizes much better over the feature space (as evidenced by the rules shown in Table 6), and therefore exhibits robustness when the rules are used on data samples that were not part of the training process.

8 Final Comments and Future Directions

While algorithm development is still at a preliminary stage, our results indicate that XCS, with appropriate pre-processing, can be a powerful tool for learning accurate, general rules that can detect and classify vehicles in hyperspectral imagery. Of greater significance is our proposed methodology for an end-to-end ATR process that relies on fusing information from different spectra. While current results show proof-of-concept on a single scene, our future efforts will focus on developing algorithms that show robustness in a number of different conditions. We believe that with further development, XCS can become part of a higher-level, general-purpose ATR system that operates in conjunction with a human ATR specialist. Of particular interest for future work is the integration of temporal data, pre-processed to include behavioral cues in moving image data. Given our success in fusing spatial data with data from multiple spectra, we believe that XCS can effectively incorporate this additional data in a functional ATR system.

Acknowledgements

The authors gratefully acknowledge support provided by the United States Air Force through FA8750-06-C-0101. They also gratefully acknowledge The Rochester Institute of Technology's (RIT) Center for Imaging Science for providing the WASP data used in this work. In particular, the authors would like to acknowledge discussions with Dr. John Kerekes of RIT that were critical in identifying and obtaining appropriate datasets for this effort.

References

1. Back, T., Fogel, D. B., Michalewicz, Z. (Eds.) (1997). *The Handbook of Evolutionary Computation*, Oxford University Press, New York
2. Butz, M., Wilson, S. W. (2002). An algorithmic description of XCS. *Soft Computing*, **6**(3–4), 144–153
3. De Jong, K. A., Spears, W. M., Gordon, D. F. (1993). Using genetic algorithms for concept learning. *Machine Learning*, **13**, 161–188
4. ftp://ftp-illigal.ge.uiuc.edue/pub/src/XCS/XCS1.2.tar.Z
5. Goldberg, D. E. (1989). *Genetic Algorithms in Search, Optimization, and Machine Learning*, Addison Wesley, MA
6. Landgrebe, D. A. (2003). *Signal Theory Methods in Multispectral Remote Sensing*, Wiley, New York

7. Lanzi, P. L., Stolzmann, W., Wilson, S. W. (Eds.) (2000). *Learning Classifier Systems: From Foundations to Applications*, Springer, Berlin Heidelberg New York, pp. 63–81
8. Malik, J., Belongie, S., Leung, T. K., Shi, J. (2001). Contour and texture analysis for image segmentation. *International Journal of Computer Vision*, **43**(1), 7–27
9. *Office of Geographic and Environmental Information (MassGIS)*, Commonwealth of Massachusetts, Executive Office of Environmental Affairs
10. Pugh, M., Waxman, A., Fay, D. (2006). Assessment of multi-sensor neural image fusion and fused data mining for land cover classification. In *Proceedings of the International Conference on Information Fusion*, Florence, Italy, July 2006
11. Ravichandran, B., Gandhe, A., Smith, R. E. Machine learning for robust automatic target recognition. *Information Fusion: The International Journal on Multi-Sensor, Multi-Source Information Fusion* (in press)
12. Ravichandran, B., Gandhe, A., Smith, R. E. (2005). XCS for robust automatic target recognition. In *Proceedings of GECCO 2005*, ACM Press, Washington DC, pp. 1803–1810
13. Smith, R. E., Dike, B. A., Ravichandran, B., El-Fallah, A., Mehra, R. K. (2001). Discovering novel fighter combat maneuvers in simulation: simulating test pilot creativity. In P. J. Bentley, D. W. Corne (Eds.), *Creative Evolutionary Systems*, Morgan Kaufmann, San Francisco, CA, pp. 467–486
14. Sutton, R., Barto, A. (1998). *Reinforcement Learning*, MIT Press, Cambridge, MA
15. *WASP Fact Sheet*, Rochester Institute of Technology
16. Wilson, S. W. (1994). Classifier fitness based on accuracy, *Evolutionary Computation*, **3**(2), 149–175, 1994
17. Wilson, S. W. (1994). ZCS: a zeroth-level classifier system, *Evolutionary Computation*, **2**(1), 1–18

Foreign Exchange Trading Using a Learning Classifier System

Christopher Stone and Larry Bull

Faculty of Computing, Engineering and Mathematical Sciences, University of the West of England, Bristol, BS16 1QY, United Kingdom, christopher.stone@uwe.ac.uk, larry.bull@uwe.ac.uk

Summary. We apply a simple Learning Classifier System to a foreign exchange trading problem. The performance of the Learning Classifier System is compared to that of a Genetic Programming approach from the literature.

The simple Learning Classifier System is able to achieve a positive excess return in simulated trading, but results are not yet fully competitive because the Learning Classifier System trades too frequently. However, the Learning Classifier System approach shows potential because returns are obtained with no offline training and the technique is inherently adaptive, unlike many of the machine learning methods currently employed for financial trading.

1 Introduction

Foreign exchange (FX) trading is the buying and selling of currencies. The estimated average *daily* volume of trade in FX is $1.2 trillion [4]. This level of transaction volume makes the currency markets some fifty times larger than the equity markets. A detailed description of FX markets is given in [10]. As with all real-world financial data, FX time series are non-linear, non-stationary and noisy. The currency markets are extremely efficient and the development of trading models for FX is difficult.

Early work using Evolutionary Computation for Computational Finance applications, for example [6,28], was usually carried out using a Genetic Algorithm. However, a GA with a fixed-length binary representation restricts the search space and for this reason, Genetic Programming (GP) [17] is more often used in Computational Finance applications, for example [21,25,27], due to its ability to represent arbitrary expressions. Interest in the use of Reinforcement Learning is also growing [13,23,39].

However, to date very few investigations have been undertaken into the use of Learning Classifier Systems for financial applications. Mahfoud and Mani [22] used a Learning Classifier System for stock prediction and Beltrametti et al. [7] used one for foreign exchange prediction. Recent work

C. Stone and L. Bull: *Foreign Exchange Trading using a Learning Classifier System*, Studies in Computational Intelligence (SCI) **125**, 169–189 (2008)
www.springerlink.com

by Schulenburg and Ross [29–31] used a Learning Classifier System for trading stocks, while Armano et al. [3] used a hybrid Neural Learning Classifier System to forecast stock indices.

Financial trading is an online endeavour. Market conditions are constantly changing and techniques are required that are able to adapt to these conditions, whilst still delivering good online performance.

Financial trading is expressible naturally as a Reinforcement Learning problem, with an agent learning to map market conditions (environmental states) to trades (actions) as a result of the reward that it receives. Reinforcement Learning is particularly suitable for financial trading because it is an online learning technique that provides scalar feedback to the agent. This gives information on the magnitude of an outcome as well as information on its correctness.

A Learning Classifier System is one possible way of implementing this framework which offers the ability to generalize over environmental states and creates rules that are readily interpretable.

A common machine learning approach to financial trading is to attempt to predict the future behaviour of a financial time series and use the prediction to make trading decisions. However, trading involves making additional decisions other than simply prediction. Trading and prediction involve searches in different spaces and it is not necessarily the case that optimizing a prediction criteria is always desirable for trading applications. Bengio [8] and de la Maza and Yuret [12] argue for evaluation based on maximizing a decision criterion such as a measure of return and not a prediction criterion such as mean squared error. Learning Classifier Systems typically operate with some form of reward function[1], which matches well the notion of decision criteria for trading. Furthermore, it is possible to employ a variety of different reward functions, which provide some measure of risk-adjusted return [11,24], as demanded by the financial trading community.

Another potential benefit of the Learning Classifier System approach is that trading decisions are made from the combined weights of multiple classifiers via the action selection algorithm. This should provide robustness that is not easily achievable in traditional approaches to financial trading, such as Genetic Algorithms and Genetic Programming, where a single 'best' rule is usually used to make a trading decision.

In this chapter, we apply a simple Learning Classifier System to a foreign exchange trading problem. To provide a benchmark against which the performance of the Learning Classifier System can be measured on real foreign exchange data, we adopt the framework and data set used by Neely et al. [25].

This chapter is arranged as follows. Section 2 provides an introduction to foreign exchange trading and an approach previously reported in the literature upon which our experiments are based. Details are given in Sect. 3 of the

[1] Except where latent learning is employed.

Learning Classifier System used and the experiments performed are presented in Sect. 4. Section 5 provides conclusions to the work.

2 Foreign Exchange Trading

2.1 Technical Analysis

There are two main approaches to financial analysis - fundamental and technical.

Fundamental analysis attempts to measure the intrinsic value of a security in order to determine its price. This is done using economic metrics directly relating to the security concerned. For example, fundamental analysis for a company might include consideration of its sales, earnings, assets, debt and many other variables.

In contrast, technical analysis does not attempt to estimate the intrinsic value of a security, but instead uses statistics from historical market information, such as past prices and volume, to determine the future price of a security. Many practitioners of technical analysis (*chartists*) have historically plotted market prices as a time series and used features of the resulting charts to make market timing decisions. Such techniques are subjective and for many years technical analysis was ignored by the academic community. However, technical analysis may also be used in an objective manner (so called *mechanical systems*) and academic studies have shown merit in this approach [14, 19, 20, 38].

The basic unit of a technical trading system is an *indicator*. An indicator is a function of the time series used (typically price history) and many types of indicator are used by technical traders. Indicators are combined into expressions to generate technical trading rules of the form (for example) 'buy if the five day moving average is greater than the 20 day moving average.' We use technical trading rules in this chapter, but these are restricted to indicators in common use. Neely [26] provides a very readable introduction to technical analysis.

2.2 Foreign Exchange Transactions

A foreign exchange (FX) rate is the amount of one currency (the *base currency*) that must be sold (bought) in order to buy (sell) one unit of another currency (the *counter currency*). An FX transaction consists of borrowing an amount of one currency and using the proceeds to fund the purchase of another currency. A *long position* in the counter currency is opened when a trader borrows enough of the base currency to fund the purchase of an amount of the counter currency at the prevailing exchange rate. The position is closed when the trader sells the counter currency at the then current exchange rate and uses this amount to repay the loan of the base currency. If, in the time between the position being opened and closed, the exchange rate rises then the trader will have made a profit on the transaction.

One of the attractions of FX trading is that the market is symmetrical and provides opportunities for profit in all market conditions. A *short position* may be taken if the trader believes that the exchange rate will fall. In this transaction the trader borrows an amount of the counter currency and uses this to fund the purchase of the base currency.

Most FX traders must operate a *margin* account. Trader's funds are held on deposit in the margin account and are used to allow risk management on the part of the institution providing dealing facilities.

Though most FX trading is carried out at high frequency, to simplify matters and to allow comparison with work appearing in the literature, we focus here on daily FX data. This is FX data where the quoted price is obtained by sampling once per trading day at a fixed time of day. In the scenario considered, a trader may, in addition to profit arising from changes in exchange rates, also benefit from any differential in the interest rate in force from positions held in the respective currencies.

2.3 Transaction Costs and Slippage

To be effective, a trading agent model must take account of *transaction costs* and *slippage* incurred in executing a trade. Transaction costs are the commission that is paid for execution of the trade and the difference between the bid (selling) price of a security and its asking (purchase) price. Slippage is the difference between the estimated and actual transaction costs. These costs are measured in terms of *basis points*. A basis point is one hundredth of a percent of the cost of the transaction. Typical transaction costs for FX trading are in the order of 2–4 basis points. Early studies of trading models typically ignored the effects of transaction costs and slippage. They often concluded that it was possible to make significant profit using these trading models. However, these conclusions are erroneous, since transaction costs and slippage have a large effect on the ability of a model to trade profitably [13].

2.4 Genetic Programming Approach

Neely et al. [25] used Genetic Programming (GP) [17] to evolve technical trading rules. This Genetic Programming approach built on work originally performed by Allen and Karjalainen [2], which, though published after that of Neely et al., was available earlier as a working paper. Much of the attraction for generating trading rules using GP is due to its ability to build arbitrary mathematical functions from a predefined set of primitive operators. In principle, this allows technical trading rules to be generated and tested with relatively little bias on the part of the researcher or practitioner. Previous work in the literature did not usually have this ability and typically studied small fixed sets of technical trading rules.

Table 1. Genetic Programming approach with validation period 1/1/1981–30/9/1995

	USD/GBP	USD/DEM	USD/CHF	USD/JPY	DEM/JPY	GBP/CHF
APR	2.28%	6.05%	1.42%	2.34%	4.10%	1.02%
Monthly std dev	3.66%	3.49%	3.88%	3.48%	2.79%	2.92%
Rules +ve return	85%	96%	84%	65%	85%	89%
Sharpe Ratio	0.18	0.50	0.11	0.19	0.42	0.10
Number of trades	130.51	106.54	156.58	107.98	426.61	55.25
Long positions	63.42%	50.52%	81.73%	78.01%	49.91%	93.57%

The data set used in this study consisted of six sets of financial time series, each containing daily data from January 1, 1974 to October 11, 1995. The time series contain data for the United States Dollar (USD), German Deutsche Mark (DEM), British Pound (GBP), Japanese Yen (JPY) and the Swiss Franc (CHF). Each data set consisted of three related time series: the exchange rate for the day concerned and the interest rates in force on each day for the base and counter currencies. Each time series contained around 5,260 elements. Data from the year 1974 was used to provide a buildup period for the indicators used, as these require history information to operate. Further details of the time series are given in [25].

The procedure used for evolving rules was to evaluate a population of 500 initially random rules using data from 1975–1977 (the *training period*). The best performing rule from the training period was saved. A new generation of 500 rules was then created from the initial population using recombination[2] and the best performing rule of that generation was compared over a *selection period* of data from 1978–1980 to the best performing rule saved previously. The better rule was retained. This procedure was repeated for 50 generations or terminated earlier if no new best rule emerged after 25 generations. If the resulting single rule produced a negative excess return (see Sect. 3.2) over the selection period it was discarded. Otherwise, its out of sample performance was determined on a *validation period* of data from 1981–1995. The study presented the results of 100 such rules, which are shown in Table 1. A detailed commentary of the results is given in Sect. 4.2.

3 Learning Classifier System Approach

3.1 Representation

There are two main differences to consider when adapting the Genetic Programming approach to a Learning Classifier System.

[2] The authors do not mention the use of any form of mutation operator for the generation of new rules.

Firstly, an attraction of using a Learning Classifier System is to exploit the fact that it is an adaptive online learning technique. Although it would be possible to perform experiments in batch mode, as was done with the Genetic Programming approach, we would ultimately like to use the Learning Classifier System to facilitate online trading, so we will emphasize this aspect in our experiments.

The second difference arises from the fact that we do not use s-expressions. Although the representation we use offers some of the expressivity of s-expressions without the issues arising from a variable-length representation, it does not offer the full degree of expression provided by Genetic Programming. In particular, we are not able to describe arbitrary mathematical expressions with the continuous-valued representation used. It is important to note that this is a direct consequence of the choice of interval representation and is not an inherent limitation of a Learning Classifier System. The use of Genetic Programming s-expression representations have been investigated with Learning Classifier Systems in [1, 18].

Trading indicators are based on four primitive functions of the time series. These are the same functions used in [25]. These primitive functions operate on elements of the exchange rate time series from the current day, t back in time to $t - \Delta t$. $lag(t, \Delta t)$ returns the exchange rate at day $t - \Delta t$. $avg(t, \Delta t)$, $min(t, \Delta t)$ and $max(t, \Delta t)$ return the mean, minimum and maximum value, respectively, of the exchange rate from day $t - \Delta t$ to day t. Though the set of primitive functions is quite small they allow the construction of popular technical indicators such as moving average, filter and trading range breakout rules.

To provide a reasonable choice of technical trading indicators within the constraints of an interval representation all indicators are of the form:

$$I_t = log\left(\frac{f_1(t, \Delta t_1)}{f_2(t, \Delta t_2)}\right) \tag{1}$$

$$f_1, f_2 \in \{lag, avg, min, max\}$$

$$0 \leq \Delta t_1 < \Delta t_2$$

$$0 \leq \Delta t_2 < 255$$

In (1), an indicator I_t for day t is the ratio of two primitive functions of the time series, f_1 and f_2. The primitive function comprising the numerator of the ratio is constrained to operate on a smaller time window $\Delta t_1 < \Delta t_2$. This halves the search space by avoiding redundant searches for equivalent indicators where one indicator is the reciprocal of another.

Indicators are created by the cover operator with the two primitive functions being chosen at random and the integer parameter Δ_t for the primitive functions being chosen uniform randomly from the appropriate range. These parameters are encoded into a single interval predicate $[\Delta t_1, \Delta t_2]$ using an 8-bit binary encoding. This technique is a convenient means of using the same interval representation for integers that we have used for real numbers. When

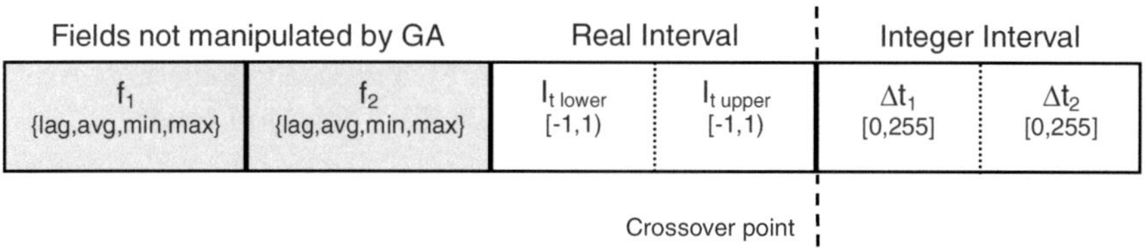

Fig. 1. Fields comprising the condition part of a classifier. For clarity, fields are shown not encoded

using a binary encoding for numbers, the encoding for integers differs from that of real numbers only in the size of the alphabet used and the consequent size of the genotypic search space.

Although this interval exists as part of the classifier's condition, its sole use is to describe an aspect of the trading rule and it is not used for matching. The Genetic Programming approach uses a 250-day history period for its trading rules, whereas the 8-bit encoding used for the Learning Classifier System allows parameters to be in the range [0, 255]. Rather than having to deal with the additional complexity of out of range values we employ a 255-day history period.

A second, continuous-valued, interval predicate, is used to define the range that the indicator I_t may take for matching to occur. This interval is in the range $[-1, 1)$. In the Genetic Programming approach, foreign exchange time series values are normalized by dividing each raw value by a 250-day moving average. With indicators being constrained to the form shown in (1) such normalization is not necessary since the indicators used are ratios that stay within certain empirical bounds. Figure 1 shows the fields that comprise a classifier's condition.

For example, a log indicator $lag(4)/max(10)$ with a range of $[0.032, 0.457)$ and an action of one would state the rule 'go long if the exchange rate four days ago is greater than 1.033 to 1.579 times the maximum exchange rate of the past ten days' and the log indicator $min(11)/avg(50)$ with a range of $[-0.112, 0.859)$ and an action of -1 would state the rule 'go short if the minimum exchange rate over the past 11 days is 0.894 to 2.361 times the mean exchange rate over the last 50 days'.

Genetic search occurs on the two interval predicates representing an indicator's range and Δt parameters. To preserve the ordering semantics of the Δt parameters and to maintain continuity with previous work, crossover *between* predicates is used. Crossover occurs such that primitive functions and their parameters are exchanged. For example, if $lag(4)/max(10)$ and $min(11)/avg(50)$ are crossed, the indicators that result are $lag(11)/max(50)$ and $min(4)/avg(10)$. As there are only two interval predicates, crossover between the condition and action parts of a classifier would, if allowed, change the action of a classifier with a high probability. For this reason, we restrict crossover to occur only in the condition part of a classifier. Mutation is used to vary the range and Δt parameters and the classifier's action in the normal manner. We have not yet experimented with extending the genetic search to include more flexible forms of variation of primitive function types.

3.2 Excess Return

Each classifier advocates one of two possible actions, long (1) or short (−1), indicating the direction of the position recommended by the classifier. One trial occurs for each element t of the time series and the Learning Classifier System adopts for trial t the position chosen by its action selection mechanism. When the Learning Classifier System receives time series data for trial $t+1$ it is able to calculate the additional return obtained by trading compared to that which would result from interest accrued on the margin. This is known as *excess return*.

The excess return r_t for a trial is influenced by the exchange rate S at trial t and $t+1$ and the differential between the daily interest rate i_{bt} for the base currency in force at trial t and that of the counter currency i_{ct}. As the data sets contain only data for days when the markets were trading, a position open over a weekend or holiday will accrue interest for $\Delta t_p > 1$ days. The trading model adopted assumes that all of the agent's margin is traded on each trial.

For a long position, excess return is calculated according to the following formula:

$$r_t = \frac{S_{t+1}}{S_t} \frac{(1+i_{ct})^{\Delta t_p}}{(1+i_{bt})^{\Delta t_p}} \tag{2}$$

For a short position, excess return is:

$$r_t = 2 - \frac{S_{t+1}}{S_t} \frac{(1+i_{ct})^{\Delta t_p}}{(1+i_{bt})^{\Delta t_p}} \tag{3}$$

Excess return is calculated for both the position adopted and the position that was not chosen for a particular trial. From this information the correctness of the position is ascertained and classifiers are allocated to the correct and incorrect sets (see [36]) to receive the relevant reward.

In a single-step model, such as that used here, the Learning Classifier System is rewarded based on the return received for a single transaction. The obvious choice is to arrange for the reward the Learning Classifier System receives to be related to the return generated by the action of the Learning Classifier System. This is a natural approach since the profit made from an action matches well the notion of utility used in many Learning Classifier Systems. The advantage of this technique is that it allows rules that generate numerically higher returns to be distinguished from other rules.

The probability distribution of financial returns maps the size of a financial return to its probability of occurrence. The nature of these distributions is much studied and debated by financial researchers, but in general terms it is known that the distribution of FX returns is fat-tailed, approximately symmetric and with a mean of almost zero [10, Chap. 5].

The Learning Classifier System is rewarded with the value $10^6(r_t - 1)$ according to the excess return obtained. For continuity with earlier work, this value is designed to provide rewards of similar magnitude to those used in [36] for stochastic test problems. Reward received is shared amongst members of the correct set.

The effects of transaction costs mean that it is possible for the correct action to receive a negative excess return. This disrupts the fitness values of the Learning Classifier System and is incompatible with the operation of the Learning Classifier System's GA. To solve this problem, reward must be limited to positive values only or additional mechanisms must be added to the Learning Classifier System to prevent a rule's fitness becoming less than zero. We adopt the former solution and to provide appropriate reinforcement the correct set is rewarded with a fixed reward of 1,000 in this case.

3.3 Transaction Costs

In Sect. 2.3 it was noted that any real-world trading activity involves some form of transaction costs. In the trading model, the agent is constrained to be always in the market and there is both a position opened and a position closed each time the current position is changed. Equation 4 models this using a charge c_p to cover the cost of changing position. This proportional cost is deducted from the excess return figure for each position.

$$c_p = \frac{(1-c)}{(1+c)} \tag{4}$$

For all trials $c = 0.00025$ (2.5 basis points). This value is the same as that used in the Genetic Programming approach in the validation period. Neely et al. [25] use a higher transaction cost of $c = 0.0005$ for training and selection to bias the genetic search towards rules that trade infrequently. However, as the focus of the present work is for online trading where there are no training, selection and validation periods, we use the same transaction cost for all trials.

4 Experiments

4.1 Experimental Setup

The Learning Classifier System that we use is based on ZCS [41] with a continuous-valued Unordered Bound Representation [35]. As described in Sect. 3, an 8-bit binary encoding is used for real numbers and integers. To improve performance and robustness on single-step Boolean problems, we use a simplified version of the NewBoole [9] update mechanism instead of the normal ZCS update algorithm.

NewBoole is a strength-based Learning Classifier System with many similarities to ZCS. This is not entirely unexpected since ZCS and NewBoole share a common predecessor, Boole [40]. One notable difference between NewBoole and ZCS is the update algorithm. NewBoole is specifically designed for problems with a Boolean action space and has an update algorithm tailored to such problems. Unlike ZCS, which rewards the action set A, NewBoole partitions the match set M into a correct set C and an incorrect set $C' = M - C$. The correct set contains all classifiers cl in the match set that advocated the correct action. This can only be computed after reward has been received and the validity of the Learning Classifier System's action has been determined. Thus, the NewBoole update is a form of supervised learning as incoming scalar reward is mapped to a notion of correctness that is independent of the action taken. This is in contrast to the standard ZCS update, which operates solely on levels of reward without classifying these in any way and is therefore a reinforcement learning approach.

The NewBoole update has a 'symmetrical payoff-penalty' strategy whereby correct and incorrect actions are updated without regard to which particular action was chosen by the performance component of the Learning Classifier System. The NewBoole update detailed in [9] uses information on the generality of a classifier as part of the update process. Following the ZCS spirit of simplicity, we ignore this aspect of the update and use a simplified version of the procedure. Specifically, we set NewBoole's 'generality emphasis' parameter G to zero. This provides resource sharing identical to that used in the standard ZCS update. NewBoole also distributes the full reward received to the correct set, rather than a fraction controlled by the learning rate β, as happens in ZCS. However, this algorithmic difference does not affect operation of the update and we retain the Widrow–Hoff scheme used by ZCS. In fact, when $\beta = 1$ resource sharing is identical in the two algorithms. Equations 5 and 6 detail the modified NewBoole update procedure we use for ZCS. For consistency, we adopt the same terminology for the learning rate parameter β and the tax rate parameter τ used in ZCS rather than the NewBoole nomenclature.

$$str \leftarrow (1-\beta)str + \beta\frac{rwd}{|C|} \;\forall\; cl \in C \tag{5}$$

$$str \leftarrow (1-\tau)str \;\forall\; cl \in C' \tag{6}$$

The NewBoole update when used with fitness proportionate (roulette wheel) action selection, as is the case in ZCS, matches closely the RPS learning rule described in [15]. Harley shows this to be an evolutionarily stable (ES) learning rule, a rule for learning evolutionarily stable strategies [34].

We have also changed the covering algorithm and use a new *specialize* operator. The parameterless cover operator that we use removes the need for the cover spread parameter s_0, which is known to be highly sensitive [35], and the specialize operator helps combat over general classifiers. Details of these changes to ZCS are given in [36].

The parameter set for the enhanced Learning Classifier System that we are using consists of just six parameters: population size limit N, GA invocation rate ρ, crossover rate χ, mutation rate μ, mutation range m and learning rate β.

Several settings have been removed from the ZCS parameter set. The initial fitness assigned to classifiers introduced by cover, f_I, does not appear to affect system performance and we set $f_I = 0$ so that new classifiers are treated no differently to classifiers that have been through one or more update cycles. Cover is hardly ever invoked because of a weak match set and we have removed this heuristic from the algorithm by setting $\phi = 0$. As expected, this makes no difference to system performance. The tax rate, τ is fixed at one, causing incorrect classifiers to receive zero fitness and effectively removing this parameter.

Another important omission from the parameter set is that of $P\#$, the amount of generality introduced into the population by the cover operator. Though this parameter is used in ternary representations, it is not strictly required with a continuous-valued representation, such as that used in the present work. The continuous-valued equivalent to the '#' symbol is the maximally general interval. It would be possible to construct a cover operator that provided a controllable proportion of maximally general intervals and to a degree this is what continuous-valued cover operators do with settings of $s_0 = [0.5, 1]$. However, we wish to reduce the number of problem-dependent system parameters. $P\#$ is certainly highly problem specific and we have obviated the need for specific settings of both s_0 and $P\#$ with the use of a parameterless cover operator and the specialize operator.

Running the enhanced Learning Classifier System on a number of difficult continuous-valued test problems ([36]) shows that the new operators and parameter settings are effective with no detrimental side-effects on any of the problems tested.

The use of artificial test problems also allowed us to develop the Learning Classifier System without recourse to financial data and any consequent biases introduced from overuse of small financial data sets. Data snooping, where data is used multiple times for the purposes of statistical model building and evaluation, is a well-known problem and challenge in data mining and machine learning [16] and also for technical trading [37].

As the population size limit N and GA invocation rate ρ are problem dependent, we ran some experiments using the USD/GBP data to determine suitable settings for these parameters. We evaluated the effects of these two parameter settings using a risk-adjusted performance measure over the entire data set and across 100 runs. This metric was the mean total excess return divided by the standard deviation of total excess return across runs. These tests yielded strong indications that a high GA rate relative to the population size limit was needed and that the GA rate required increased with population size limit. Since each classifier's condition consists of only two intervals a large population is not necessary and we use $N = 50$ and $\rho = 4^3$ for all experiments using the enhanced Learning Classifier System.

In our parameter setting experiments we found that because the number of elements in the time series was very low compared to the problem complexity, it was advantageous for the cover operator to introduce multiple classifiers into the population. This allows the GA to test multiple hypotheses arising from a single environmental situation and provides the diversity necessary for the GA to function effectively. As a result, in these experiments the cover operator introduces eight classifiers into the population each time it is invoked.

In earlier work, we did not investigate the effect of the GA crossover rate χ, mutation rate μ and mutation range m, so these parameters were left at the values used for previous experiments, $\chi = 0.5$, $\mu = 0.002$, $m = 0.1$. The learning rate was also set to the previously used value of $\beta = 0.2$.

One hundred runs were performed of each experiment. Results are shown for all six currency pairs, but as the USD/GBP currency pair was used to set the value of two of the Learning Classifier System parameters, results for this currency pair may not be statistically valid for the reasons of data snooping described above.

4.2 Validation Period from 1981 to 1995

We first restrict the period under which the Learning Classifier System is evaluated to 1981–1995. This matches the validation period used for the Genetic Programming approach. In this scenario, the Learning Classifier System learns using data from 1975 through to 1995, but the performance of the Learning Classifier System is only measured during the period 1981–1995. Table 2 shows the performance of the Learning Classifier System using the architecture developed in [35] with the parameter settings detailed in Sect. 4.1. For comparison, Table 1 shows the results published in [25] achieved by the Genetic Programming approach.

Table 2. Enhanced Learning Classifier System with validation period 1/1/1981–30/9/1995

	USD/GBP	USD/DEM	USD/CHF	USD/JPY	DEM/JPY	GBP/CHF
APR	3.86%	0.34%	2.80%	0.86%	3.58%	−1.23%
Monthly std dev	3.44%	3.49%	3.70%	3.35%	2.75%	2.70%
Runs +ve return	94%	48%	82%	56%	96%	20%
Sharpe Ratio	0.33	0.06	0.25	0.11	0.4	−0.12
Number of trades	627.84	648.68	659.21	663.2	589.7	638.45
Long positions	52.52%	50.77%	49.34%	46.33%	47.66%	44.61%

[3] That is, four sets of parents producing offspring per cycle.

The first line of the tables shows the Annual Percentage Rate excess return achieved during the validation period 1981–1995. The results show that the Learning Classifier System achieved a positive excess return on five out of six currency pairs. In contrast, the Genetic Programming approach achieved positive returns on all six currency pairs. This is largely attributable to the fact that in the Genetic Programming approach rules achieving a negative excess return are discarded during the selection period. This is evidenced by the percentage of rules in the Genetic Programming approach that have positive returns, which is high in all cases.

In the Genetic Programming approach one fixed rule is used per run, whereas in the Learning Classifier System approach, 50 competing adaptive rules are used per run. However, the Learning Classifier System selects only one action per trial, so we can compare the results of 100 Genetic Programming rules with 100 runs of the Learning Classifier System. The number of Learning Classifier System runs with positive excess returns correlates well with the mean excess return achieved.

Excess return in itself is insufficient to judge performance. The risk taken by a trading agent in achieving profit must also be taken into account. Measurement of risk involves consideration of volatility of returns. A simple way of doing this is to measure the monthly standard deviation of excess return over non-overlapping periods. This measure is comparable for both Genetic Programming and Learning Classifier System approaches across all currency pairs and is explained in [25] as being because rules are required to be in the market at all times with a long or short position, which have similar variances.

$$\begin{aligned} D_t &= R_{Ft} - R_{Bt} \\ \bar{D} &= \frac{1}{T}\sum_{t=1}^{T} D_t \\ S_h &= \frac{\bar{D}}{\sigma_D} \end{aligned} \tag{7}$$

A metric encountered often in the finance literature is the Sharpe Ratio [32,33]. There are several forms of the Sharpe Ratio. Here, we use the ex post Sharpe Ratio shown in (7). This measures the mean differential return $\bar{D}$ achieved from the trading strategy R_F compared to that of a benchmark R_B divided by the standard deviation σ_D of differential returns over the period. In this case the benchmark R_B is the risk-free interest i received on the margin, that would accrue from a strategy of not trading. Returns are measured for non-overlapping periods of length t over a sample period T. Here, we measure annual returns over the validation period.

Sharpe Ratios for the Learning Classifier System approach reflect the distribution of returns, with a greater spread across the currencies than that seen with the Genetic Programming approach. Results show that the Learning Classifier System approach can achieve a higher Sharpe Ratio than the Genetic Programming approach, but that it is also possible to achieve poorer

Table 3. Enhanced Learning Classifier System with exploit trials in validation period 1/1/1981–30/9/1995

	USD/GBP	USD/DEM	USD/CHF	USD/JPY	DEM/JPY	GBP/CHF
APR	−3.05%	−3.60%	−2.60%	−0.85%	−0.19%	−2.32%
Monthly std dev	3.47%	3.46%	3.79%	3.32%	2.80%	2.75%
Runs +ve return	4%	2%	4%	41%	65%	2%
Sharpe Ratio	−0.18	−0.22	−0.11	−0.05	0.01	−0.22
Number of trades	849.55	674.91	610.85	949.33	490.79	489.53
Long positions	72.81%	16.04%	13.37%	79.45%	89.02%	12.09%

results. This may be due to the number of trades executed by the Learning Classifier System. For all currency pairs the number of trades is around 600, several times that of the Genetic Programming approach, which shows considerably more variability in trading frequency.

The number of long positions taken by the Learning Classifier System stays quite close to 50%. On the other hand, the Genetic Programming approach again shows more variability in the number of long positions. However, the Genetic Programming approach uses only a single fixed rule for each run, whereas the Learning Classifier System uses multiple adaptive rules, so we may expect that this effect would occur in an efficient market.

To determine the effects of adaptation on the results obtained, we repeated the experiment with no GA or update activity taking place during the validation period. In this situation, the Learning Classifier System trades using the fixed set of rules that were in force at the start of the validation period unless covering is necessary. To gauge the risk of the cover operator disrupting the population with random classifiers, we checked the frequency of covering during a single run of the Learning Classifier System and found only one occurrence of cover throughout the validation period.

The results of this experiment are presented in Table 3. These show that the Learning Classifier System produces a negative excess return for all currency pairs and that the number of runs with positive returns is much lower than when GA and update activity occurs. The number of long positions taken diverges considerably from 50% and shows significant variation across currencies, as does the number of trades, reflecting the composition of rules in force at the start of the validation period. However, such a snapshot of rules clearly do not provide a good trading performance over the validation period. This result supports the hypothesis that, during the validation period, adaptation of rules is occurring within the Learning Classifier System which is of benefit to the agent's performance.

4.3 Validation Period from 1975 to 1995

In the previous section, we used a validation period from 1981 to 1995 to allow a meaningful comparison of the Learning Classifier System and Genetic

Table 4. Enhanced Learning Classifier System with validation period 1/2/1975–30/9/1995

	USD/GBP	USD/DEM	USD/CHF	USD/JPY	DEM/JPY	GBP/CHF
APR	4.14%	0.67%	3.48%	2.92%	4.97%	0.75%
Monthly std dev	3.23%	3.31%	3.67%	3.25%	2.87%	2.92%
Runs +ve return	99%	61%	94%	89%	100%	67%
Sharpe Ratio	0.37	0.09	0.31	0.29	0.51	0.11
Number of trades	845.45	896.63	902.84	896.7	820.47	881.85
Long positions	56.00%	51.94%	49.52%	46.90%	47.02%	44.67%

Programming approaches. To do this we compared the performance of 100 runs of the Learning Classifier System against that of 100 rules produced by Genetic Programming.

One significant difference between the two approaches is the number of evaluations occurring for each rule or run. In the Genetic Programming approach, the number of evaluations needed to create and test the 100 rules used is in the region 1,000–2,000 million. In contrast, the number of evaluations used for the 100 runs of the Learning Classifier System is in the region of only 26 million!

Of course, such comparisons are irrelevant provided that the elapsed time needed to create and test rules is less than the frequency with which new rules are needed and the rules provide some utility. Rather, it is informative to realize how much the Learning Classifier System is able to achieve when observing each daily price only once per run. Because the amount of data is rather small for efficient induction, and because the agent is intended to be used online, we now present results over the entire data set. That is, the validation period is extended to cover the period 1975[4]–1995 and the agent is now evaluated on every trading decision it makes during its lifetime.

Table 4 shows the results for the enhanced Learning Classifier System architecture. These correspond to results presented in Table 2 for the shorter validation period. Over the entire data set, the Learning Classifier System achieves a positive excess return on all currency pairs. Furthermore, for all currency pairs, excess return, Sharpe Ratio and number of runs with a positive excess return are higher than for the shorter validation period. Further testing is necessary to determine if this improved performance is due to market conditions or the larger number of trials available to the Learning Classifier System.

To assess whether the Learning Classifier System developments detailed in [36] indeed have any bearing on the results obtained, we ran a standard version of ZCS with the baseline parameters used in earlier work. These were $N = 50$, $\beta = 0.2$, $\tau = 0.1$, $\chi = 0.5$, $\mu = 0.002$, $m = 0.1$, $f_I = 20$, $s_0 = 0.25$, $\rho = 0.25$,

[4] The validation period actually begins on February 1, 1975 to allow for a 255-day buildup period at the start of each data set.

Table 5. ZCS with standard settings and validation period 1/2/1975–30/9/1995

	USD/GBP	USD/DEM	USD/CHF	USD/JPY	DEM/JPY	GBP/CHF
APR	1.29%	0.00%	0.00%	−0.20%	−1.13%	0.03%
Monthly std dev	3.43%	3.40%	3.89%	3.52%	3.09%	3.13%
Runs +ve return	97%	46%	6%	47%	33%	8%
Sharpe Ratio	0.14	0.03	0.03	0.02	−0.06	0.02
Number of trades	11.07	6.92	6.41	6.16	8.27	6.34
Long positions	93.20%	49.03%	71.49%	48.92%	42.43%	47.86%

$\phi = 0.5$. Only a single classifier was introduced with each invocation of the cover operator. Results are shown in Table 5. They show poor performance compared to that seen in Table 4. This is due to the extremely low number of trades made by ZCS.

Investigation of individual runs of this experiment revealed that lack of trading was caused by the population being taken over by classifiers advocating the same action. This is caused by the following sequence of events: Classifiers created by cover tend to be quite general and therefore take part in many match sets. Each time a classifier matches it gets taxed either by being promoted to the action set and paying a fraction $\beta = 0.2$ of its fitness into the bucket or by not making the action set and paying a tax $\tau = 0.1$ of its fitness. Because classifiers participate in many match sets, they must receive reward relatively often in order to avoid their fitness decreasing. If this does not occur, for example when a classifier makes a few incorrect predictions, fitness decreases quite rapidly. Under roulette wheel action selection, these classifiers are less likely to be selected for the action set and they become increasingly likely to remain in the match set, from which they cannot obtain reward. In this way a gap opens between the relative fitness of classifiers advocating opposing actions.

The problem is compounded by the low frequency of invocation of the GA relative to the frequency of matching and consequent fitness decay. Without frequent GA activity, it is possible for classifiers to have their fitness reduced to almost zero. When the GA does operate, the roulette selection mechanism used favours selection of the stronger action. If two classifiers selected for reproduction both have the same action, crossover will produce offspring with that action, which can only be changed with low probability by mutation. Under the GA, the stronger action is propagated within the population. Within a relatively few trials it is possible for all of the classifiers suggesting the weaker action to have such a low fitness that there is little realistic chance of them reproducing. Over a longer time period and within the 5,000 or so trials of these data sets, replacement seals the fate of the weaker species and they are totally eradicated from the population which then consists solely of classifiers advocating the same action.

We repeated the experiment with a setting of $s_0 = 0.05$ to cause the cover operator to introduce narrower intervals into the population. Though trading frequency increased marginally, the change did not prevent population takeover by classifiers of the same action. It is of note that this situation cannot occur with the NewBoole update since the entire match set is updated on every trial, not just members of the action set as is the case with the ZCS update algorithm.

To provide further control experiments we tested versions of the enhanced Learning Classifier System used to provide the results in Table 4 with the exception that the position chosen was random or a fixed short or fixed long position. The results of these experiments are shown in Tables 6 to 8 respectively.

Table 6. Enhanced Learning Classifier System with random position and validation period 1/2/1975–30/9/1995

	USD/GBP	USD/DEM	USD/CHF	USD/JPY	DEM/JPY	GBP/CHF
APR	−6.77%	−6.02%	−5.89%	−6.26%	−6.13%	−6.21%
Monthly std dev	3.10%	3.10%	3.57%	2.93%	2.57%	2.75%
Runs +ve return	0%	0%	0%	0%	0%	0%
Sharpe Ratio	−0.67	−0.59	−0.5	−0.65	−0.72	−0.67
Number of trades	2621.02	2615.07	2618.78	2625.63	2618.65	2619.91
Long positions	50.00%	50.00%	50.02%	49.94%	50.03%	49.98%

Table 7. Enhanced Learning Classifier System with fixed short position and validation period 1/2/1975–30/9/1995

	USD/GBP	USD/DEM	USD/CHF	USD/JPY	DEM/JPY	GBP/CHF
APR	−1.63%	−1.75%	−1.24%	−4.82%	−2.91%	−0.14%
Monthly std dev	3.43%	3.41%	3.89%	3.48%	3.08%	3.11%
Runs +ve return	0%	0%	0%	0%	0%	0%
Sharpe Ratio	−0.03	−0.05	0.01	−0.28	−0.18	0.06
Number of trades	2	2	2	2	2	2
Long positions	0.00%	0.00%	0.00%	0.00%	0.00%	0.00%

Table 8. Enhanced Learning Classifier System with fixed long position and validation period 1/2/1975–30/9/1995

	USD/GBP	USD/DEM	USD/CHF	USD/JPY	DEM/JPY	GBP/CHF
APR	0.48%	0.59%	−0.28%	3.98%	2.17%	−0.78%
Monthly std dev	3.43%	3.39%	3.89%	3.53%	3.10%	3.15%
Runs +ve return	100%	100%	0%	100%	100%	0%
Sharpe Ratio	0.1	0.1	0.04	0.34	0.24	−0.02
Number of trades	2	2	2	2	2	2
Long positions	100.00%	100.00%	100.00%	100.00%	100.00%	100.00%

A control strategy of choosing positions at random performs especially badly. This is no doubt due at least in part to the large number of trades and consequent transaction costs incurred in trading. The fixed short position provides negative excess return on all runs and is clearly a poor strategy. However, the fixed long position does surprisingly well for a naive strategy and provides a positive excess return on four of the currency pairs. Even so, the enhanced Learning Classifier System exceeds the performance of a fixed long position on five of the six currency pairs measured in terms of excess return and on four of the currency pairs in terms of Sharpe Ratio.

5 Conclusions

We have shown that a simple Learning Classifier System architecture, properly configured, is able to achieve a positive excess return in simulated trading. Although results are not yet fully competitive with those obtained from Genetic Programming, the Learning Classifier System approach is attractive because these positive returns have been obtained without any offline training. Furthermore, the technique is inherently adaptive unlike many of the machine learning methods currently employed for financial trading.

Trading frequency of the Learning Classifier System approach was much higher than that seen with Genetic Programming and this is almost certainly the factor limiting performance compared to the Genetic Programming approach. One area worthy of further study is to see whether trading frequency can be lowered to reduce the effect of transaction costs on the returns achieved. It would appear that this may be difficult using a single-step model as is used in the present work. This is because with a single-step model successive trials are independent and there is no direct method for the Learning Classifier System to control trading frequency, which is a function of the actions chosen in successive trials. To achieve any further control over trading frequency, it may be necessary to extend the agent architecture to include additional state information or to model the trading activity as a multi-step problem.

The returns in this chapter have been achieved using historical data. The availability of information and communications technologies that were not available until recently means that such returns may be difficult or impossible to achieve in live trading. For this reason we do not claim that the architecture or parameter settings used are in any way optimal or final. Rather we see the results presented here as providing evidence of the usefulness and potential of Learning Classifier Systems for financial trading. In this respect, these results echo those of Schulenburg and Ross [29–31], who also reported encouraging results from a simple Learning Classifier System architecture.

In the interests of comparing the Learning Classifier approach with that of Genetic Programming, we adopted a similar experimental setup. In particular, we provided the same sensory information to the agent as was used for the Genetic Programming approach, namely the exchange rate time series.

Interest rate information was used only to calculate excess return and was not available to the agent for use when selecting a position. Yet, the differential between the interest rates received for the base and counter currencies forms an important part of the overall excess return achieved by the agent. It is quite possible for there to be a large interest rate differential between the two currencies in a pair, which could dominate a trading decision, but with the experimental setup currently in use the agent does not have this information. It would seem that including such interest rate information as additional sensory input would be beneficial to performance.

Similarly, for transaction costs and slippage to be included in a trading agent's model of the environment, the environment should include a sensor representing the agent's current position. Without these changes the environment is partially observable, since the same apparent environmental state may result in differing rewards depending on the current interest rate differential and the position held by the agent in the previous time step.

Finally, the Learning Classifier System approach is well suited to enhancement through an ensemble approach, which should aid performance. The low computational cost of an individual Learning Classifier System means that it is realistic to run many individual agents in parallel and still achieve high throughput. This is an important consideration when using high frequency data, as is necessary for a realistic trading model where near real-time decisions must be made. These topics, however, are beyond the scope of the present work and will be the subject of future developments.

Acknowledgements

The authors are grateful to Christopher Neely for kindly making available the data used for the experiments.

References

1. M. Ahluwalia and L. Bull. A Genetic Programming-Based Classifier System. In Banzhaf et al. [5], pages 11–18.
2. F. Allen and R. Karjalainen. Using Genetic Algorithms to find technical trading rules. *Journal of Financial Economics*, 51(2):245–271, 1999.
3. G. Armano, M. Marchesi, and A. Murru. NXCS: Hybrid approach to stock indexes forecasting. In S. -H. Chen, editor, *Genetic Algorithms and Genetic Programming in Computational Finance*, pages 125–158. Kluwer, Boston, MA, 2002.
4. Bank for International Settlements. *Triennial Central Bank Survey. Foreign Exchange and Derivatives Market Activity in 2001.* Document from the Monetary and Economic Department, Basel, Switzerland, 2002.
5. W. Banzhaf, J. Daida, A. E. Eiben, M. H. Garzon, V. Honavar, M. Jakiela, and R. E. Smith, editors, *GECCO-99: Proceedings of the Genetic and Evolutionary Computation Conference*, Morgan Kaufmann, San Francisco, 1999.
6. R. J. Bauer, Jr. *Genetic Algorithms and Investment Strategies.* Wiley, New York, 1994.

7. L. Beltrametti, R. Fiorentini, L. Marengo, and R. Tamborini. A learning-to-forecast experiment on the foreign exchange market with a Classifier System. *Journal of Economic Dynamics and Control*, 21(8–9):1543–1575, 1997.
8. Y. Bengio. Using a financial training criterion rather than a prediction criterion. *International Journal of Neural Systems*, 8(4):433–443, 1997.
9. P. Bonelli, A. Parodi, S. Sen, and S. W. Wilson. NEWBOOLE: A fast GBML system. In B. W. Porter and R. J. Mooney, editors, *Proceedings of the Seventh International Conference on Machine Learning*, pages 153–159. Morgan Kaufmann, San Mateo, 1990.
10. M. M. Dacorogna, R. Gençay, U. Müller, R. B. Olsen, and O. V. Pictet. *An Introduction to High-Frequency Finance*. Academic, San Diego, 2001.
11. M. M. Dacorogna, R. Gençay, U. A. Müller, and O. V. Pictet. Effective return, risk aversion and drawdowns. *Physica A*, 289(1–2):229–248, 2001.
12. M. de la Maza and D. Yuret. A critique of the standard Neural Network application to financial time series analysis. http://citeseer.ist.psu.edu/146873.html.
13. M. A. H. Dempster, T. W. Payne, and Y. S. Romahi. Intraday FX trading: Reinforcement vs evolutionary learning. Judge Institute of Management Working Paper 23/2001, University of Cambridge, Cambridge, UK, 2001.
14. R. Gençay. Linear, non-linear and essential foreign exchange rate prediction with simple technical trading rules. *Journal of International Economics*, 47(1): 91–107, 1999.
15. C. B. Harley. Learning the evolutionarily stable strategy. *Journal of Theorectical Biology*, 89(4):611–633, 1981.
16. D. Jensen. Data snooping, dredging and fishing: The dark side of data mining a SIGKDD99 panel report. *SIGKDD Explorations*, 1(2):52–54, 2000.
17. J. R. Koza. *Genetic Programming*. MIT, Cambridge, 1992.
18. P. L. Lanzi. Extending the Representation of Classifier Conditions, Part ii: From Messy Coding to S-Expressions. In Banzhaf et al. [5], pages 345–352.
19. B. LeBaron. Do moving average trading rule results imply non-linearities in foreign exchange markets? Working Paper 9222, University of Wisconsin-Madison, Social Systems Research Institute, 1992.
20. R. M. Levich and L. R. Thomas, III. The significance of technical trading-rule profits in the foreign exchange market: A bootstrap approach. *Journal of International Money and Finance*, 12(5):451–474, 1993.
21. J. Li and E. P. K. Tsang. Investment decision making using FGP: A case study. In P. J. Angeline, Z. Michalewicz, M. Schoenauer, X. Yao, and A. Zalzala, editors, *Proceedings of the Congress of Evolutionary Computation*, pages 1253–1259. IEEE Press, Piscataway, 1999.
22. S. Mahfoud and G. Mani. Financial forecasting using Genetic Algorithms. *Applied Artificial Intelligence*, 10(6):543–565, 1996.
23. J. Moody and M. Saffell. Learning to trade via direct reinforcement. *IEEE Transactions on Neural Networks*, 12(4):875–889, 2001.
24. J. Moody and L. Wu. Optimization of trading systems and portfolios. In A. S. Weigend, Y. Abu-Mostafa, and A.-P. N. Refenes, editors, *Decision Technologies for Financial Engineering*, volume 7 of *Progress in Neural Processing*, pages 23–35. World Scientific, Singapore, 1997.
25. C. Neely, P. Weller, and R. Dittmar. Is technical analysis in the foreign exchange market profitable? a Genetic Programming approach. *Journal of Financial and Quantitative Analysis*, 32(4):405–427, 1997.

26. C. J. Neely. Technical analysis in the foreign exchange market: A layman's guide. *Federal Reserve Bank of St. Louis Review*, pages 23–38, September/October 1997.
27. M. Oussaidène, B. Chopard, O. V. Pictet, and M. Tomassini. Parallel Genetic Programming: An application for trading model induction. In J. R. Koza, D. Goldberg, and R. L. Riolo, editors, *Genetic Programming 1996, Proceedings of the First Annual Conference*, 1996.
28. O. V. Pictet, M. M. Dacorogna, B. Chopard, M. Oudsaidène, R. Schirru, and M. Tomassini. Using Genetic Algorithms for robust optimization in financial applications. *Neural Network World*, 5(4):573–587, 1995.
29. S. Schulenburg and P. Ross. An adaptive agent based economic model. In P. L. Lanzi, W. Stolzmann, and S. W. Wilson, editors, *Learning Classifier Systems. From Foundations to Applications*, volume LNAI-1813 of *Lecture Notes in Artificial Intelligence*, pages 265–284. Springer, Berlin Heidelberg, New York, 2000.
30. S. Schulenburg and P. Ross. Strength and money: An LCS approach to increasing returns. In P. L. Lanzi, W. Stolzmann, and S. W. Wilson, editors, *Advances in Learning Classifier Systems. Proceedings of the Third International Workshop (IWLCS-2000)*, volume LNAI-1996 of *Lecture Notes in Artificial Intelligence*, pages 114–137. Springer, Berlin Heidelberg New York, 2001.
31. S. Schulenburg and P. Ross. Explorations in LCS models of stock trading. In P. L. Lanzi, W. Stolzmann, and S. W. Wilson, editors, *Advances in Learning Classifier Systems. Proceedings of the Fourth International Workshop (IWLCS-2001)*, volume LNAI-2321 of *Lecture Notes in Artificial Intelligence*, pages 151–180. Springer, Berlin Heidelberg New York, 2002.
32. W. F. Sharpe. Mutual fund performance. *Journal of Business*, 39(1):119–138, 1966.
33. W. F. Sharpe. The Sharpe Ratio. *Journal of Portfolio Management*, 21(1):49–58, 1994.
34. J. Maynard Smith. *Evolution and the Theory of Games*. Cambridge University Press, Cambridge, UK, 1982.
35. C. Stone and L. Bull. For real! XCS with continuous-valued inputs. *Evolutionary Computation*, 11(3):299–336, 2003.
36. C. Stone and L. Bull. Configuring ZCS for continuous-valued single-step Boolean problems. Technical Report UWELCSG05-006, University of the West of England Learning Classifier Group, Bristol, UK, 2005. http://www.cems.uwe.ac.uk/lcsg/reports/uwelcsg05-006.ps.
37. R. Sullivan, A. Timmermann, and H. White. Data-snooping, technical trading rule performance and the bootstrap. Disussion Paper 97–31, Department of Economics, University of California, San Diego, 1997.
38. M. P. Taylor and H. Allen. The use of technical analysis in the foreign exchange market. *Journal of International Money and Finance*, 11(3):304–314, 1992.
39. B. Van Roy. Temporal-difference learning and applications in finance. In Y. S. Abu-Mostafa, B. LeBaron, A. W. Lo, and A. S. Weigend, editors, *Computational Finance 1999*, pages 447–461. MIT, Cambridge, 1999.
40. S. W. Wilson. Classifier systems and the animat problem. *Machine Learning*, 2(3):199–228, 1987.
41. S. W. Wilson. ZCS: A zeroth order classifier system. *Evolutionary Computation*, 2(1):1–18, 1994.

Towards Clustering with Learning Classifier Systems

Kreangsak Tamee[1,2], Larry Bull[2], and Ouen Pinngern[1]

[1] Department of Computer Engineering, Faculty of Engineering, Research Center for Communication and Information Technology (ReCCIT), King Mongkut's Institute of Technology Ladkrabang, Bangkok 10520, Thailand, kreangsakt@yahoo.com, kpouen@kmitl.ac.th

[2] School of Computer Science, University of the West of England, Bristol BS16 1QY, UK, larry.bull@uwe.ac.uk

Summary. This chapter presents a novel approach to clustering using an accuracy-based Learning Classifier System. Our approach achieves this by exploiting the generalization mechanisms inherent to such systems. The purpose of the work is to develop an approach to learning rules which accurately describe clusters without prior assumptions as to their number within a given dataset. Favourable comparisons to the commonly used k-means algorithm are demonstrated on a number of synthetic datasets.

1 Introduction

This chapter presents initial results from a rule-based approach to clustering through the development of an accuracy-based Learning Classifier System (LCS) [10]. A number of studies have indicated good performance for LCS in classification tasks (e.g., see [2] for examples). We are interested in the utility of such systems to perform unsupervised learning tasks.

Clustering is an important unsupervised learning technique where a set of data are grouped into clusters in such a way that data in the same cluster are similar in some sense and data in different clusters are dissimilar in the same sense. For this it is necessary to first define a measure of similarity which will establish a rule for assigning data to the domain of a particular cluster centre. One such measure of similarity may be the Euclidean distance D between two data x and y defined by $D = ||x - y||$. Typically in data clustering there is no one perfect clustering solution of a dataset, but algorithms that seek to minimize the cluster spread, i.e., the family of centre-based clustering algorithms, are the most widely used (e.g., [23]). They each have their own mathematical objective function which defines how well a given clustering solution fits a given dataset. In this paper our system is compared to the

K. Tamee et al.: *Towards Clustering with Learning Classifier Systems*, Studies in Computational Intelligence (SCI) **125**, 191–204 (2008)
www.springerlink.com

most well-known of such approaches, the k-means algorithm. We use as a measure of the quality of each clustering solution the total of the k-means objective function:

$$o(X, C) = \sum_{i=1}^{n} \min_{j \in \{1 \ldots k\}} ||x_i - c_j||^2 \tag{1}$$

Define a d-dimensional set of n data points $X = \{x_1, \ldots, x_n\}$ as the data to be clustered and k centers $C = \{c_1, \ldots, c_k\}$ as the clustering solution. However most clustering algorithms require the user to provide the number of clusters (k), and the user in general has no idea about the number of clusters (e.g., see [16]). Hence this typically results in the need to make several clustering trials with different values for k where $k = 2$ to $k_{max} =$ square-root of n (data points) and select the best clustering among the partitioning with different number of clusters. The commonly applied Davies–Bouldin [6] validity index is typically used as a guideline to the underlying number of clusters here.

Previously, evolutionary algorithms have been used for clustering in two principle ways. The first uses them to search for appropriate centers of clusters with established clustering algorithms such as the k-means algorithm, e.g., the GA-clustering algorithm [11]. However this approach typically requires the user to provide the number of clusters. Tseng and Yang [17] proposed the CLUSTERING algorithm which has two stages. In the first stage a nearest-neighbor algorithm is used to reduce the size of data set and in the second the GA-clustering algorithm approach is used. Sarafis [13] has recently proposed a further stage which uses a density-based merging operator to combine adjacent rules to identify the underlying clusters in the data. We suggest that modern accuracy-based LCS are well-suited to the clustering problem due to their generalization capabilities.

The chapter is structured as follows: first we describe the general scheme for using accuracy-based LCS for clustering and then present initial results. The adoption of a more sophisticated fitness function is found to be beneficial. A form of rule compaction for clustering with LCS, as opposed to classification, is then presented. A form of local search is then introduced before a number of increasingly difficult synthetic datasets are used to test the algorithm.

2 A Simple LCS for Clustering

In this chapter we begin by presenting a version of the simple accuracy-based YCS [3] which is derived from XCS [18], here termed YCSc. YCSc is a Learning Classifier System without internal memory, where the rulebase consists of a number (N) of rules. Associated with each rule is a scalar which indicates the average error (ε) in the rule's matching process and an estimate of the average size of the niches (match sets – see below) in which that rule participates (σ). The initial random population of rules have their parameters set to 10.

On receipt of an input data, the rulebase is scanned, and any rule whose condition matches the message at each position is tagged as a member of the current match set [M]. The rule representation here is the Centre-Spread encoding (see [14] for discussions). A condition consists of interval predicates of the form $\{\{c_1, s_1\}, \ldots.. \{c_d, s_d\}\}$, where c is the interval's range centre from [0.0,1.0] and s is the "spread" from that centre from the range $(0.0, \mathrm{s_0}]$ and d is a number of dimensions. Each interval predicates' upper and lower bounds are calculated as follows: $[c_i - s_i, c_i + s_i]$. If an interval predicate goes outside the problem space bounds, it is truncated. A rule matches an input x with attributes x_i if and only if $c_i - s_i \leq x_i < c_i + s_i$ for all x_i.

Reinforcement in YCSc consists of updating the matching error ε which is derived from the Euclidean distance with respect to the input x and c in the condition of each member of the current [M] using the Widrow-Hoff delta rule with learning rate β:

$$\varepsilon_j \leftarrow \varepsilon_j + \beta((\sum_{l=1}^{d} (x_l - c_{lj})^2)^{1/2} - \varepsilon_j) \tag{2}$$

Next, the niche size estimate is updated:

$$\sigma_j \leftarrow \sigma_j + \beta(|[\mathrm{M}]| - \sigma_j) \tag{3}$$

YCSc employs two discovery mechanisms, a niche genetic algorithm (GA) [9] and a covering operator. The general niche GA technique was introduced by Booker [1], who based the trigger on a number of factors including the payoff prediction "consistency" of the rules in a given [M], to improve the performance of LCS. XCS uses a time-based mechanism under which each rule maintains a time-stamp of the last system cycle upon which it was consider by the GA. The GA is applied within the current niche when the average number of system cycles since the last GA in the set is over a threshold θ_{GA}. If this condition is met, the GA time-stamp of each rule in the niche is set to the current system time, two parents are chosen according to their fitness using standard roulette-wheel selection, and their offspring are potentially crossed and mutated, before being inserted into the rulebase. This mechanism is used here within match sets, as in the original XCS algorithm [18], which was subsequently changed to work in action sets to aid generalization per action [4].

The GA uses roulette wheel selection to determine two parent rules based on the inverse of their error:

$$f_i = \frac{1}{{\varepsilon_i}^{v} + 1} \tag{4}$$

Offspring are produced via mutation (probability μ) where, after [19], we mutate an allele by adding an amount + or − $rand(m_0)$, where m_0 is a fixed real, rand picks a real number uniform randomly from $(0.0, m_O]$, and the sign is chosen uniform randomly. Crossover (probability χ, two-point) can occur between any two alleles, i.e., within an interval predicate as well as between predicates, inheriting the parents' parameter values or their average if crossover is invoked. Replacement of existing members of the rulebase uses roulette wheel selection based on estimated niche size. If no rules match on a given time step, then a covering operator is used which creates a rule with its condition centre on the input value and the spread with a range of $rand(s_O)$, which then replaces an existing member of the rulebase in the same way as the GA.

Recently, Butz et al. [5] have proposed a number of interacting "pressures" within XCS. Their "set pressure" considers the more frequent reproduction opportunities of more general rules. Opposing the set pressure is the pressure due to fitness since it represses the reproduction of inaccurate overgeneral rules. Thus to produce an effective, i.e., general but appropriately accurate, solution an accuracy-based LCS using a niche GA with global replacement should have these two pressures balanced through the setting of the associated parameters. In this chapter we show how the same mechanisms can be used within YCSc to identify clusters within a given dataset; the set pressure encourages the evolution of rules which cover many data points and the fitness pressure acts as a limit upon the separation of such data points, i.e., the error.

3 Initial Performance

In this section we apply YCSc as described above on two datasets for the first experiment to test the performance of the system. The first dataset is well-separated as shown in Fig. 1a. We use a randomly generated synthetic

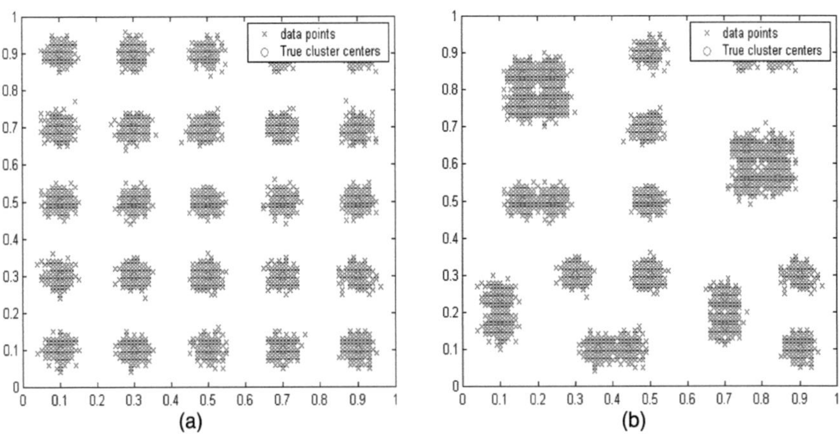

Fig. 1. The well-separated (**a**) and less-separated (**b**) data sets used

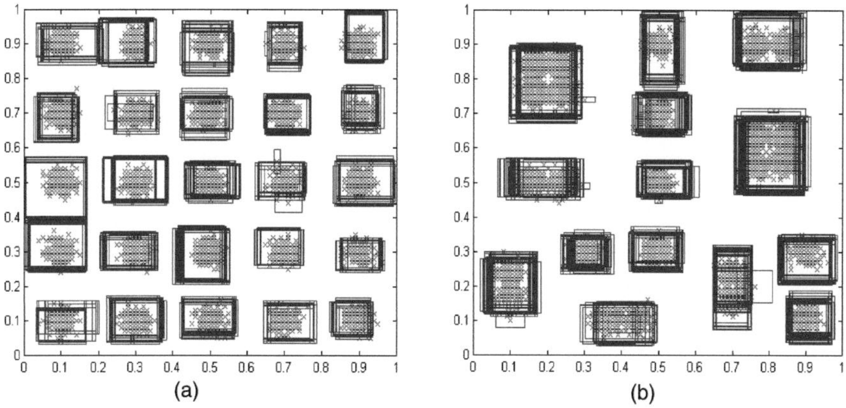

Fig. 2. Typical solutions for the well-separated (**a**) and less-separated (**b**) data sets

dataset. This dataset has $k = 25$ true clusters arranged in a 5×5 grid in $d = 2$ dimension. Each cluster is generated from 400 data points using a Gaussian distribution with a standard deviation of 0.02, for a total of $n = 10{,}000$ datum. The second dataset is not well-separated as shown in Fig. 1b. We generated it in the same way as the first dataset except the clusters are not centred on that of their given cell in the grid.

The parameters used were: $N = 800$, $\beta = 0.2$, $\upsilon = 5$, $\chi = 0.8$, $\mu = 0.04$, $\theta_{GA} = 12$, $s_0 = 0.03$, $m_0 = 0.006$. All results presented are the average of ten runs. Learning trials consisted of 200,000 presentations of a randomly sampled data point.

Figure 2 shows typical example solutions produced by YCSc on both data sets. That is, the region of the $2D$ input space covered by each rule in the final rule-base is plotted along with the data. As can be seen, in the well-separated case the system roughly identifies all 25 clusters whereas in the less-separated case contiguous clusters are covered by the same rules.

As expected, solutions contain many overlapping rules around each cluster. The next section presents a rule compaction algorithm which enables identification of the underlying clusters.

4 Rule Compaction

Wilson [20] introduced a rule compaction algorithm for XCS to aid knowledge discovery during classification problems (see also [7,8,22]). We have developed a compaction algorithm for clustering:

Step 1: Delete the useless rules: The useless rules are identified and then deleted from the ruleset in the population based on their coverage. Low coverage means that a rule matches a small fraction (20%) of the average coverage.

Step 2: Find the required rules from numerosity: The population $[\mathrm{P}]_{N[deleted]}$ is sorted according to the numerosity of the rules and delete the rules that have lower numerosity, less than 2. Then $[\mathrm{P}]_M$ $(M < N)$ is formed by selecting the minimum sequential set of rules that covers all data.

Step 3: Find the required rules from average error: The population $[\mathrm{P}]_M$ is sorted according to the average error of the rules. Then $[\mathrm{P}]_P$ $(P < M)$ is formed by selecting the minimum sequential set of rules that covers all data.

Step 4: Remove redundant rules: This step is an iterative process. On each cycle it selects the rule in $[\mathrm{P}]_P$ that maximum number of match set. This rule is removed into the final ruleset $[\mathrm{P}]_F$ and the data that it covers deleted from the dataset. The process continues until the dataset is empty.

Figure 3 shows the final set $[\mathrm{P}]_F$ for both the full solutions shown in Fig. 2. YCSc's identification of the clusters is now clear. Under the (simplistic) assumption of non-overlapping regions as described by rules in $[\mathrm{P}]_F$ it is easy to identify the clusters after compaction. In the case where no rules subsequently match new data we could of course identify a cluster by using the distance between it and the centre of each rule.

We have examined the average quality of the clustering solutions produced during the ten runs by measuring the total objective function described in (1) and checking the number of clusters defined. The average quality on the well-separated dataset is 8.12 ± 0.54 and the number of clusters is 25 ± 0. That is, it correctly identifies the number of clusters every time. The average quality on the not well-separated dataset is 24.50 ± 0.56 and the number of clusters

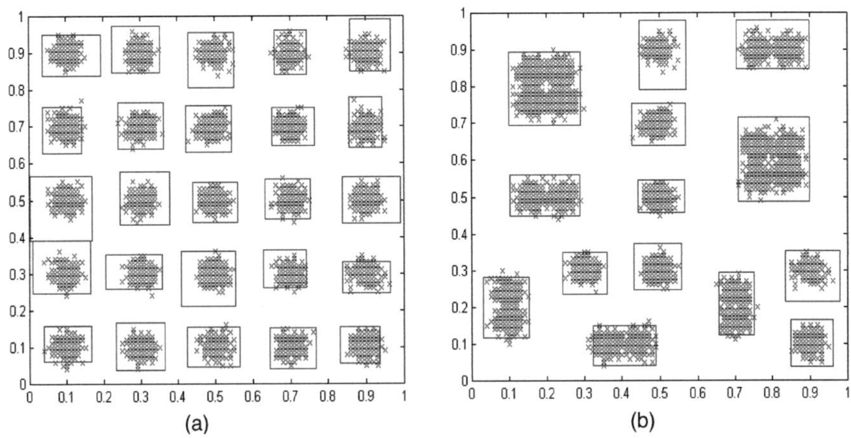

Fig. 3. Showing the effects of the rule compaction on the typical solutions shown in Fig. 2 for the well-separated (**a**) and less-separated (**b**) data sets

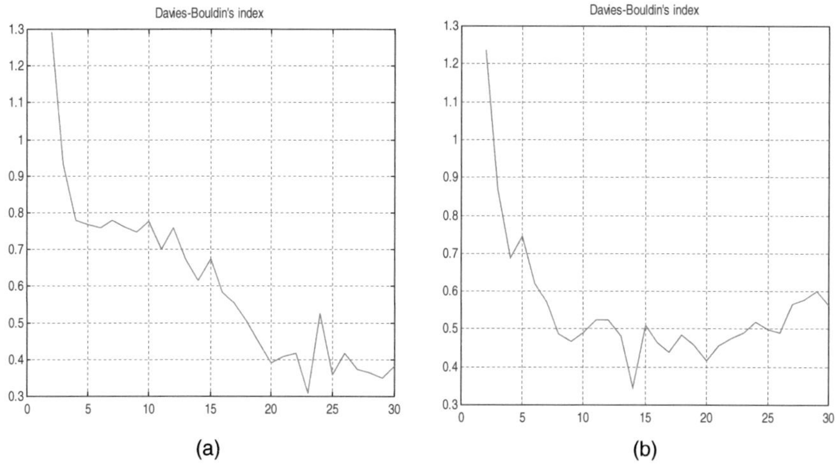

Fig. 4. k-Means algorithm performance using the Davies–Bouldin index for the well-separated (**a**) and less-separated (**b**) data sets

is 14 ± 0. Hence it is not correct every time due to the lack of clear separation in the data.

For comparison, the k-means algorithm was applied to the datasets. The k-means algorithm (assigned with the known $k = 25$ clusters) averaged over 10 runs gives a quality of 32.42 ± 9.49 and 21.07 ± 5.25 on the well-separated and less-separated datasets respectively. The low quality of solutions in the well-separated case is due to the choice of the initial centres; k-means is well-known for becoming less reliable as the number of underlying clusters increases. For estimating the number of clusters we ran, for 10 times each, different k (2–30) with different random initializations. To select the best clustering with different numbers of clusters, the Davies–Bouldin validity index is shown in Fig. 4. The result on the well-separated dataset has a lower negative peak at 23 clusters and the less-separated dataset has a lower negative peak at 14 clusters. That is, it is not correct on both datasets, for the same reason as noted above regarding quality. Thus YCSc performs as well or better than k-means whilst also identifying the number of clusters during learning.

5 Modifying XCS for Clustering

As noted above, YCS is a simplified version of XCS, presently primarily to aid understanding of how such accuracy-based LCS learn [3]. The principle difference is that fitness F is slightly more complex. First, the accuracy κ_j and the relative accuracy κ'_j are computed as

$$\kappa_j = \begin{cases} 1, \dots\dots\dots\dots\dots\dots\dots . if \varepsilon_j > \varepsilon_0 \\ \alpha\left(\frac{\varepsilon_j}{\varepsilon_0}\right)^{-\upsilon}, \dots\dots .otherwise \end{cases} \tag{5}$$

$$\kappa'_j = \frac{\kappa_j}{\sum\limits_{j=[M]} \kappa_j} \tag{6}$$

The parameter $\varepsilon_0(\varepsilon_0 > 0)$ controls the tolerance for rule error (ε); the parameter $\alpha(0 < \alpha < 1)$ and the parameter $\upsilon(\upsilon > 0)$ are constants controlling the rate of decline in accuracy κ when ε_0 is exceeded. Finally, fitness F is updated toward the current relative accuracy as follows:

$$F_j \leftarrow F_j + \beta(\kappa'_j - F_j) \tag{7}$$

The reader is referred to [4] for a full algorithmic description of XCS.

Using the same parameters as above, with $\varepsilon_0 = 0.03$ and $\alpha = 0.1$, we have examined the average quality of the clustering solutions produced during the ten runs by measuring the total objective function described in (1) and checking the number of clusters defined. The average of quality on the well-separated dataset is 6.65 ± 0.12 and the number of clusters is 25.0 ± 0. The average quality on the not well-separated dataset is 6.71 ± 0.14 and the number of clusters is 25.0 ± 0. That is, it correctly identifies the number of clusters every time. Thus XCSc performs better than both YCSc and k-means whilst also identifying the number of clusters during learning. That is, YCSc struggled with the less-separated data and analysis of solutions indicates that the difference in error between more appropriate descriptions of the underlying clusters and those typically promoted is very small, which are not sufficiently amplified under the fitness scaling of (4). The function of XCS therefore seems more appropriate for such problems (note no difference was seen for a number of classification tasks [3]).

6 Local Search

Previously, Wyatt and Bull [21] have introduced the use of local search within XCS for continuous-valued problem spaces. Within the classification domain, they used the Widrow-Hoff delta rule to adjust rule condition interval boundaries towards those of the fittest rule within each niche on each matching cycle, reporting significant improvements in performance. Here good rules serve as a basin of attraction under gradient descent search thereby complimenting the GA search. The same concept has also been applied to a neural rule representation scheme in XCS [12]. We have examined the performance of local search for clustering using Wyatt and Bull's scheme: once a focal rule (the highest fitness rule) has been identified from the current match set all rules in [M] use the Widrow-Hoff update procedure to adjust each of the two interval descriptor pairs towards those of the focal rule, e.g., $c_{ij} < -c_{ij} + \beta_l[F_j - c_{ij}]$, $\forall i, j$, where c_{ij} represents gene j of rule i in the match set, F_j represent gene j of the focal rule, and β_l is a learning set to 0.1. The spread parameters are adjusted in the same way and the mechanism is applied on every match cycle before the GA trigger is tested. Initial results using Wyatt and Bull's scheme gave a reduction in performance, typically more specific rules, i.e., too many clusters, were identified (not shown).

We here introduce a scheme which uses the current data sample as the target for the local learning to adjust only the centres of the rules:

$$c_{ij} < -c_{ij} + \beta_l(x_j - c_{ij}) \quad (8)$$

Where c_{ij} represents the centre of gene j of rule i in the current match set, x_j represents the value in dimension j of the current input data, and β_l is the learning rate, here set to 0.1. This is applied on every match cycle before the GA trigger is tested, as before. In the well-separated case, the quality of solutions was 6.50 ± 0.09. In the less-separated case, the quality of solutions was 6.48 ± 0.07. The same number of clusters was identified as before, i.e., 25 and 25 respectively. Thus results indicate that our data-driven local search improves the quality of the clustering over the non-local search approach and is used hereafter. The same was found for YCSc but it does not improve the cluster identification [15].

7 Adaptive Threshold Parameter

The ε_0 parameter controls the error threshold of rules and we have investigated the sensitivity of XCSc to its value by varying it. Experiments show that, if ε_0 is set high, e.g., 0.1, in the less-separated case the contiguous clusters are covered by the same rules (Fig. 5). We therefore developed an adaptive threshold parameter scheme which uses the average error of the current [M]:

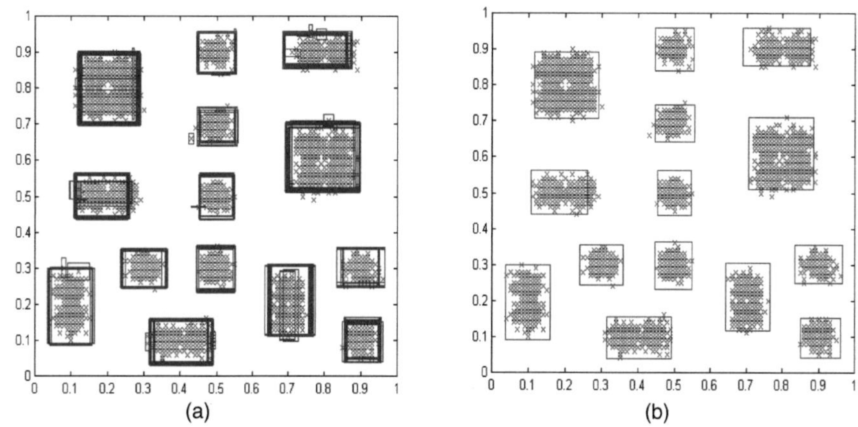

Fig. 5. Typical solutions using $\varepsilon_0 = 0.1$ before (**a**) and after (**b**) rule compaction, for the less-separated dataset

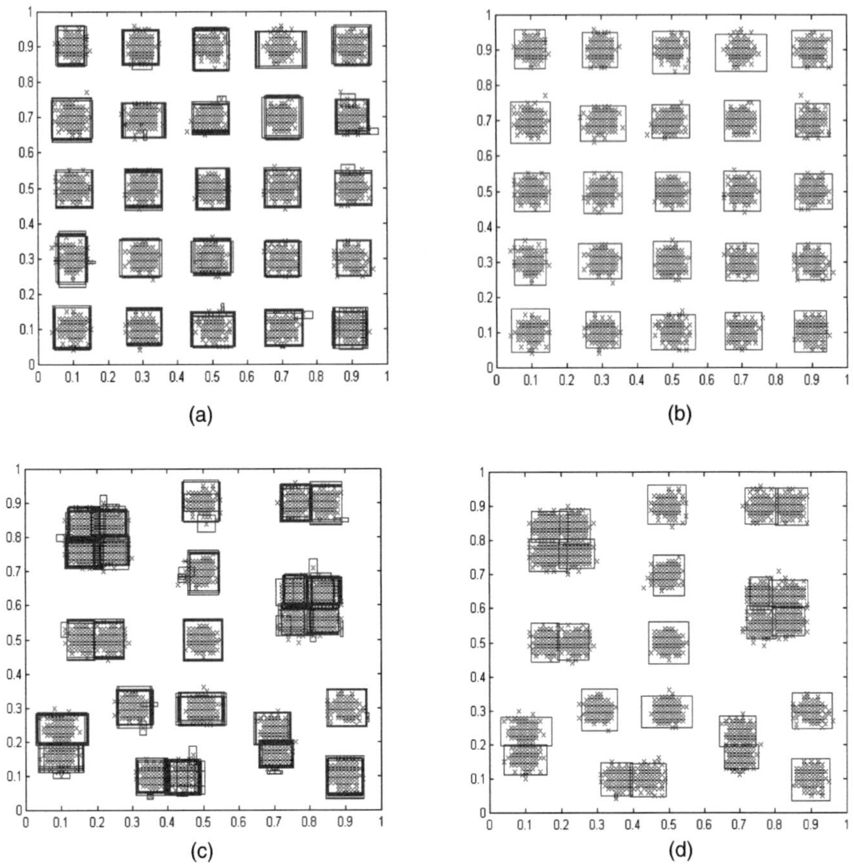

Fig. 6. Typical solutions using adaptive ε_0 approach before and after rule compaction, for well-separated (**a, b**) and less-separated (**c, d**) dataset

$$\varepsilon_0 = \tau(\sum \varepsilon_j / N_{[M]}) \tag{9}$$

Where ε_j is the average error of each rule in the current match set and $N_{[M]}$ is the number of rules in the current match set. This is applied before the fitness function calculations. Experimentally we find $\tau = 1.2$ is most effective for the problems here.

Figure 6 shows how in the well-separated case, the average quality and number of clusters from 10 runs is as before, being 6.39 ± 0.04 and 25.0 ± 0 respectively. In the less-separated case the average quality is again almost unchanged at 6.40 ± 0.09 and the number of clusters is 25.0 ± 0. There are no significant differences in average quality but with the adaptive technique there is a reduction in the number of parameters that require careful, possibly problem specific, setting by the user.

Table 1. XCSc with local search vs. k-means on harder datasets

Dataset	k-Means		XCSc	
	k Found	Quality	k Found	Quality
k = 9, d = 2	7	24.28 ± 7.63	9.00 ± 0.00	13.13 ± 0.29
k = 9, d = 4	6	83.80 ± 66.34	9.00 ± 0.00	21.94 ± 0.31
k = 9, d = 6	9	133.11 ± 44.36	9.00 ± 0.00	43.79 ± 0.23
k = 25, d = 2	24	37.37 ± 10.39	25.00 ± 0.00	18.15 ± 0.45
k = 25, d = 4	20	152.38 ± 46.94	25.00 ± 0.00	52.05 ± 0.01
k = 25, d = 6	22	278.67 ± 68.58	25.00 ± 0.00	67.78 ± 0.33

8 Increased Complexity

Here we examine the performance of XCSc compared to k-means over randomly generated datasets in several d dimensions with varying numbers of k clusters. A Gaussian distribution is generated around each centre, their standard deviation is set from 0.01 (well-separated) up to 0.05 (less-separated). Each centre coordinate is generated from a uniform distribution over the hypercube $[0, 1]^d$, the expected distances between cluster centres is set to 0.2. Thus, the expected value of the cluster separation varied inversely with standard deviation. We test datasets with d-dimensions 2, 4 and 6. The true k clusters are 9 and 25, where we generate 400 data points for each cluster.

The parameters used were as before and we determine the average quality of clustering and number of clusters from XCSc with local search from 10 runs as before. We also determine for k-means (the number of k groups was known) the quality and Davies–Bouldin index as before. Table 1 shows how XCSc always gives superior quality and gives an equivalent or closer estimate of the number of clusters compared to k-means.

We have also considered data in which the clusters are of different sizes and/or of different density, examples of which are shown in Fig. 7a and c. In both cases, using the same parameters as before, XCSc with the adaptive error threshold mechanism is able to correctly identify the true clusters, as shown in Fig. 7b and d. The system without the adaptive mechanism was unable to solve either case, neither was YCSc (not shown).

9 Conclusions

Our experiments clearly show how a new clustering technique based on the accuracy-based learning classifier system can be effective at finding clusters of high quality whilst automatically finding the number of clusters. That is, XCSc, with its more sophisticated fitness function, when adapted slightly, appears able to reliably evolve an optimal population of rules through the

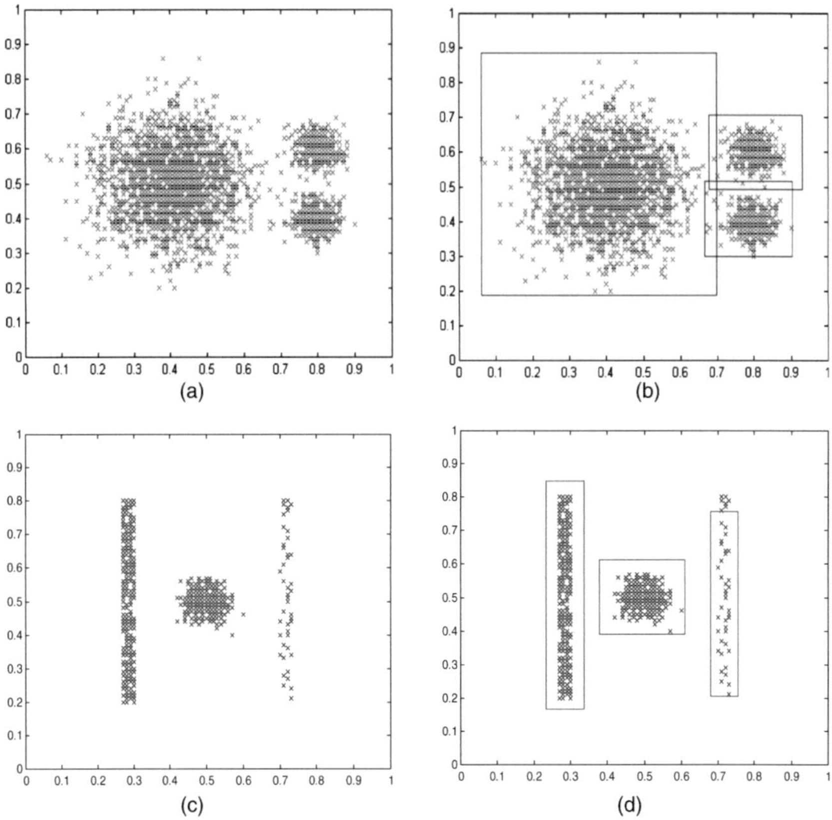

Fig. 7. Typical solutions using the adaptive ε_0 approach after rule compaction for two varingly spaced datasets

use of reinforcement learning to update rule parameters and a genetic algorithm to evolve generalizations over the space of possible clusters in a dataset. The compaction algorithm presented reduces the number of rules in the total population to identify the rules that provide the clustering. The local search mechanism helps guide the centres of the rules' intervals in the solution space to approach the true centres of clusters; results show that local search improves the quality of the clustering over a non-local search approach. As noted, the original system showed a sensitivity to the setting of the error threshold but an effective adaptive scheme has been introduced which compensates for this behaviour. We are currently applying the approach to a number of large real-world datasets and comparing the performance of XCSc to other clustering algorithms which also determine an appropriate number of clusters during learning.

References

1. Booker, L.B. (1989) Triggered Rule Discovery in Classifier Systems. In J.D. Schaffer (ed) *Proceeding of the Third International Conference on Genetic Algorithms*. Morgan Kaufmann, San Francisco, CA, pp. 265–274
2. Bull, L. (ed) (2004) *Applications of Learning Classifier Systems*. Springer, Berlin Heidelberg New York
3. Bull, L. (2005) Two Simple Learning Classifier Systems. In L. Bull & T. Kovacs (eds) *Foundations of Learning Classifier Systems*. Springer, Berlin Heidelberg New York, pp. 63–90
4. Butz, M. & Wilson, S. (2001) An Algorithmic Description of XCS. In P.L. Lanzi, W. Stolzmann, & S.W. Wilson (eds) *Advances in Learning Classifier Systems. Third International Workshop (IWLCS-2000)*, Lecture Notes in Artificial Intelligence (LNAI-1996). Springer, Berlin Heidelberg New York
5. Butz, M., Kovacs, T., Lanzi, P.-L., & Wilson, S.W. (2004) Toward a Theory of Generalization and Learning in XCS. *IEEE Transactions on Evolutionary Computation* 8(1): 28–46
6. Davies, D.L. & Bouldin, D.W. (1979) A Cluster Separation Measure. *IEEE Transactions on Pattern Analysis and Machine Intelligence* PAMI-1(2): 224–227
7. Dixon, P., Corne, D., & Oates, M. (2003) A Ruleset Reduction Algorithm for the XCS Learning Classifier System. In P.L. Lanzi, W. Stolzmann, & S. Wilson (eds) *Proceedings of the 5th International Workshop on Learning Classifier Systems*. Springer, Berlin Heidelberg New York, pp. 20–29
8. Fu, C. & Davis, L. (2002) A Modified Classifier System Compaction Algorithm. In Banzhaff et al. (eds) *Proceedings of GECCO 2002*. Morgan Kaufmann, San Francisco, CA, pp. 920–925
9. Holland, J.H. (1975) *Adaptation in Natural and Artificial Systems*. University of Michigan Press
10. Holland, J.H. (1976) Adaptation. In R. Rosen & F.M. Snell (eds) *Progress in Theoretical Biology*, vol. 4. Plenum, New York, pp. 263–293
11. Maulik, U. & Bandyopadhyay, S. (2000) Genetic Algorithm-Based Clustering Technique. *Pattern Recognition* 33: 1455–1465
12. O'Hara, T. & Bull, L. (2005) A Memetic Accuracy-based Neural Learning Classifier System. In *Proceedings of the IEEE Congress on Evolutionary Computation*. IEEE, New York, pp. 2040–2045
13. Sarafis, I.A., Trinder, P.W., & Zalzala, A.M.S. (2003) Mining Comprehensible Clustering Rules with an Evolutionary Algorithm. In E. Cant'u- Paz et al. (eds) *Proceedings of Genetic and Evolutionary Computation Conference (Gecco'03)*, LNCS 2724, pp. 2301–2312
14. Stone, C. & Bull, L. (2003) For Real! XCS with Continuous-Valued Inputs. *Evolutionary Computation* 11(3): 299–336
15. Tamee, K., Bull, L., & Pinngern, O. (2006) A Learning Classifier System Approach to Clustering. In *Sixth International Conference on Intelligent System Design and Application (ISDA), Jinan, China*. IEEE, New York, vol. ISDA I, pp. 621–626
16. Tibshirani, R., Walther, G., & Hastie, T. (2000) Estimating the Number of Clusters in a Dataset via the Gap Statistic. *Journal of the Royal Statistical Society B* 63: 411–423
17. Tseng, L.Y. & Yang, S.B. (2001) A Genetic Approach to the Automatic Clustering Problem. *Pattern Recognition* 34: 415–424

18. Wilson, S.W. (1995) Classifier Fitness Based on Accuracy. *Evolutionary Computation* 3(2): 149–176
19. Wilson, S.W. (2000) Get Real! XCS with Continuous-Valued Inputs. In P.L. Lanzi, W. Stolzmann, & S.W. Wilson (eds) *Learning Classifier Systems From Foundations to Applications.* Springer, Berlin Heidelberg New York, pp. 209–219
20. Wilson, S. (2002). Compact Rulesets from XCSI. In P.L. Lanzi, W. Stolzmann, & S.W. Wilson (eds) *Proceedings of the 4th International Workshop on Learning Classifier Systems.* Springer, Berlin Heidelberg New York, pp. 197–210
21. Wyatt, D. & Bull, L. (2004) A Memetic Learning Classifier System for Describing Continuous-Valued Problem Spaces. In N. Krasnagor, W. Hart, & J. Smith (eds) *Recent Advances in Memetic Algorithms.* Springer, Berlin Heidelberg New York, pp. 355–396
22. Wyatt, D., Bull, L., & Parmee, I. (2004) Building Compact Rulesets for Describing Continuous-Valued Problem Spaces Using a Learning Classifier System. In I. Parmee (ed) *Adaptive Computing in Design and Manufacture VI.* Springer, Berlin Heidelberg New York, pp. 235–248
23. Xu, R. & Winch, D. (2005) Survey of Clustering Algorithms. *IEEE Transactions on Neural Networks* 16(3): 645–678

A Comparative Study of Several Genetic-Based Supervised Learning Systems

Albert Orriols-Puig[1], Jorge Casillas[2], and Ester Bernadó-Mansilla[1]

[1] Enginyeria i Arquitectura La Salle, Universitat Ramon Llull, 08022, Barcelona, Spain, aorriols@salleURL.edu, esterb@salleURL.edu
[2] Department of Computer Science and Artificial Intelligence, University of Granada, 18071, Granada, Spain, casillas@decsai.ugr.es

Summary. This chapter gives insight in the use of Genetic-Based Machine Learning (GBML) for supervised tasks. Five GBML systems which represent different learning methodologies and knowledge representations in the GBML paradigm are selected for the analysis: UCS, GAssist, SLAVE, Fuzzy AdaBoost, and Fuzzy LogitBoost. UCS and GAssist are based on a non-fuzzy representation, while SLAVE, Fuzzy AdaBoost, and Fuzzy LogitBoost use a linguistic fuzzy representation. The models evolved by these five systems are compared in terms of performance and interpretability to the models created by six highly-used non-evolutionary learners. Experimental observations highlight the suitability of GBML systems for classification tasks. Moreover, the analysis points out which systems should be used depending on whether the user prefers to maximize the accuracy or the interpretability of the models.

1 Introduction

Genetic-Based Machine Learning (GBML) [30] gathers a wide range of learning techniques that use *Evolutionary Algorithms* (EAs) [5,26,30] for knowledge discovery. Research on this topic was historically conducted from two perspectives: Pittsburgh-style [12] and Michigan-style [30] GBML systems. Recently, the increase in the understanding of how evolutionary algorithms work [27] has propelled the research on GBML. As a result, the concept of GBML has been extended with new proposals of learners that use EAs to evolve their knowledge. Some of these approaches lay between the definitions of Pittsburgh-style and Michigan-style GBML, such as the *Iterative Learning Rule* approach [47]. Others methodologies propose to include EAs as robust search mechanisms to assist the building of neural networks [34,48] or statistical classifiers [14,36].

Among the different new approaches, evolutionary algorithms have been combined with *Fuzzy Rule-Based Systems*, resulting in the so-called *Genetic Fuzzy Rule-Based Systems* (GFRBSs) [11]. GFRBSs use EAs either to learn fuzzy rules or to tune different components of the fuzzy system. The main

A. Orriols-Puig et al.: *A Comparative Study of Several Genetic-Based Supervised Learning Systems*, Studies in Computational Intelligence (SCI) **125**, 205–230 (2008)
www.springerlink.com

advantage of the fuzzy representation is that it allows for a better interpretability of the knowledge evolved, providing a flexible, robust, and powerful methodology to deal with *noisy*, *imprecise*, and *incomplete* data.

The aim of this work is to evaluate the performance and the interpretability of the models evolved by five different GBML architectures in data mining tasks, and to compare their behavior to other non-evolutionary techniques. We include representatives of the different tendencies of GBML for supervised learning in the comparison. We select two GBML systems that use a non-fuzzy (*crisp*) representation: UCS [6], and GAssist [4]; and three GBML methods based on a fuzzy representation: SLAVE [29], Fuzzy AdaBoost [14], and Fuzzy LogitBoost [36]. We compare these GBML systems to six highly-used non-evolutionary techniques. These learners come from different learning paradigms such as instance-based learning, rule and decision-tree induction, statistical modeling, and neural networks. The algorithms are compared on a collection of 20 real-world datasets extracted from the UCI repository [8] and local repositories [7].

The remaining of this paper is organized as follows. Section 2 presents the different approaches in GBML, and Sect. 3 briefly explains the five GBML systems selected for the comparison. Section 4 details the experimentation comparison and present the results. Finally, Sect. 5 summarizes and concludes the work.

2 Genetic-Based Machine Learning

Firstly designed regarding the animal behavior [30], research on GBML has been historically conducted from two different perspectives: the *Pittsburgh* approach [12], and the *Michigan* approach [30, 31]. Recently, a third methodology has received an increasing amount of attention: the *Incremental Rule Learning* approach [47]. This three families are briefly introduced as follows.

Pittsburgh-Style GBML Systems follow the essence of genetic algorithms [27, 30]. Each individual in the population consists of a rule set that covers all the input space. The quality of an individual is estimated considering different aspects such as its accuracy and size. The typical genetic operators, i.e., selection, crossover, and mutation, are adapted to deal with rule sets instead of binary strings. At the end of the learning process, Pittsburgh-style GBML systems return the best individual found during the search, which is used to predict the class of unknown examples. The first successful developments of Pittsburgh-style GBML for supervised learning are GABIL [13] and GIL [32]. A new generation Pittsburgh-style GBML derived from GABIL can be found in GAssist [4].

Michigan-Style GBML Methods are *cognitive systems* that combine a credit-apportionment system, usually adapted from a reinforcement learning method [44], and evolutionary algorithms to evolve a population of accurate rules.

Differently from the Pittsburgh approach, each individual is represented by a single rule, which is evaluated incrementally by the credit-apportionment system. An evolutionary algorithm is applied with a certain frequency on the population to discover new promising rules. At the end of the learning process, all the rules in the population contribute to decide the output of new test examples. Some of the first developments of Michigan-style GBML are SCS [26] and NewBoole [9]. Although these systems showed to be able to solve different classification tasks, some drawbacks were also detected, mainly associated to the achievement of accurate generalizations. This led to further research that culminated in the design of XCS [50, 51], by far the most influential Michigan-style GBML system. XCS works under a reinforcement learning scheme. The system can be used for classification tasks by considering a reinforcement learning problem in which maximum payoffs correspond to correct classifications and minimum payoffs to incorrect classifications. On the other hand, classification tasks can be solved in a more straightforward way using UCS [6], a system which inherits the main components from XCS but specializes them to supervised learning.

Iterative Rule Learning (IRL), firstly proposed in the context of the *SIA* system [47], uses a separate-and-conquer methodology to create an ordered list of individuals. Each individual is represented by a single rule, as in the Michigan approach. The system iteratively invokes an evolutionary algorithm, which evaluates the individuals according to their accuracy and generality. The best individual returned by the EA is added to the end of the ordered list of rules, and all the matching examples are removed from the training dataset. This process is repeated until the training dataset is empty. In test mode, the predicted class of a new example is given by the first rule in the ordered list that matches the example. One of the best representatives of IRL system is HIDER [1].

The big advances in the understanding of evolutionary algorithms and the proposal of the first *competent evolutionary algorithms* [27, 37, 38] has led to the hybridization of EAs and different machine learning techniques, resulting in a wide variety of new GBML systems. In the field of statistical learning, EAs have been applied to discover new promising rules in different boosting algorithms such as AdaBoost [21] and LogitBoost [22]. In the realm of neural networks, EAs have been used to evolve either the weights [34] or the structure of neural networks [48]. The flexibility provided by EAs has also been used to construct Fuzzy Rule-Based Systems (FRBSs), resulting in the so-called GFRBS [11]. Research on this field has mainly focused on the use of evolutionary algorithms to tune different components of an FRBS, such as the fuzzy sets or the fuzzy rules.

For the study performed in this paper, we selected five rule-based GBML systems that belong to different approaches and use either fuzzy or crisp knowledge representations. Specifically, we chose two of the most significant crisp GBML methods for supervised learning, which use intervalar representations: (a) UCS [6], a Michigan-style GBML system, and (b) GAssist [4], a Pittsburgh-style GBML technique. We also selected three GBML algorithms that use a fuzzy-rule representation: (c) SLAVE [10, 28], an IRL algorithm, and the boosting techniques (d) Fuzzy AdaBoost [14] and (e) Fuzzy LogiBoost [36], two implementations of the AdaBoost and LogitBoost algorithms that use EAs to evolve fuzzy rules. Next section introduces these five learners.

3 Description the GBML Systems Used in the Comparison

This section briefly describes the five GBML systems included in the comparison: UCS [6], GAssist [4], SLAVE [29], Fuzzy AdaBoost [14], and Fuzzy LogitBoost [36]. The reader is referred to the original papers for further information about these methods.

3.1 UCS

UCS [6] is a Learning Classifier System which inherits the main components of XCS [50,51], but specializes them to be applied only to supervised learning tasks. In the following, we describe the main components of the system.

Knowledge Representation UCS evolves a population of classifiers which jointly cover the input space. Each classifier consists of a production rule of the form *condition* → *class* and a set of parameters. The condition specifies the set of inputs where the classifier can be applied. For continuous inputs, the condition is codified as a set of intervals $[l_i, u_i]^n$, which globally represents a hyperrectangle in the feature space. The class c^k of the rule specifies the class predicted when the condition is satisfied:

$$\textbf{if } x_1 \in [l_1, u_1] \wedge \ldots \wedge x_n \in [l_n, u_n] \textbf{ then } c^k \tag{1}$$

Each rule has the following parameters: a) accuracy acc; b) fitness F; c) correct set size cs; d) numerosity num; and e) experience exp. Accuracy and fitness are measures of the quality of the classifier. The correct set size is the estimated average size of all the correct sets where the classifier participates. Numerosity is the number of copies of the classifier, and experience is the number of times that a classifier has belonged to a match set.

Learning Interaction During training, UCS incrementally evolves a set of classifiers. At each learning iteration, the system receives an input example e and its class c. Then, the system creates the match set [M], which contains all the classifiers in the population [P] whose condition matches e. From that, the correct set [C] is formed, which consists of all the classifiers in [M] that predict class c. If [C] is empty, the covering operator is activated, creating a new classifier with a generalized condition matching e, and predicting class c.

Next, the parameters of all the classifiers in [M] are updated. The experience of each classifier is increased, and its accuracy is updated depending on whether the current prediction was correct. The correct set size cs is calculated if the classifier belongs to [C]. Then, as proposed in [35], the fitness is shared among all the classifiers that participate in [C].

After one learning cycle, a genetic algorithm (GA) is triggered if the average time since the last application of the GA on the classifiers in [C] is greater than θ_{GA}. In this case, the GA selects two parents from [C] with a probability that depends on the classifier's fitness. The two parents are copied, creating two new children, which are recombined and mutated with probabilities χ and μ respectively. Recombination crosses the parent's conditions by two points. Mutation modifies the lower and upper bound of an interval according to a uniform distribution. Finally, each offspring is introduced into the population, removing another classifier if the population is full.

3.2 GAssist

GAssist [4] is one of the most competitive Pittsburgh-style GBML systems. GAssist was initially derived from GABIL [13], introducing several modifications that able the system to overcome scalability problems detected in the first Pittsburgh-style GBML approaches [20]. The rule representation and the learning interaction are described as follows.

Knowledge Representation

GAssist evolves a set of individuals, each of them represented by a variable-length set of rules:

$$Ind = (R_1 \vee R_2 \vee \ldots R_n) \tag{2}$$

where each rule consists of a condition and a predicted class c^k:

$$\textbf{IF } (x_1 = V_1^1 \vee \ldots \vee x_1 = V_m^1) \wedge \ldots \wedge (x_n = V_n^1 \wedge \ldots \wedge x_n = V_n^k) \textbf{ THEN } c^k \tag{3}$$

That is, each input variable x_i is represented by a disjunction of feasible values for this variable. For nominal variables, $(V_1^i ... V_j^i)$ represent the j possible values that the variable can take. For continuous variables, GAssist applies a discretization technique to transform the input space into intervals of values. Several discretization techniques have been proposed for GAssist. In our experiments, we used a uniform discretization.

Learning Interaction

The core of the system is a near-standard genetic algorithm similar to the one applied in GABIL [13]. Thus, at each learning iteration, the system selects a set of individuals, and applies crossover and mutation generating a new set of offspring. These offspring are evaluated by means of a fitness function based on the *minimum description length principle* (MDL) [41]. GAssist uses the same crossover operator defined by GABIL, i.e., a *semantically correct crossover* operator [13]. This is a multiple-point crossover operator that forces that the selected points cut both parents in the same position of the variable. The mutation operator randomly adds or removes one value of a given variable.

GAssist introduces a new *deletion operator* that permits to remove rules from individuals, and so, to control their size. This operator is activated after a predefined number of iterations, and it removes the rules of an individual that do not match any input example. To avoid an excessive loss of diversity, which may have a negative effect in subsequent recombinations, this operator is not applied if the individual does not have a minimum number of rules.

Finally, GAssist controls the runtime of the system by means of a windowing scheme addressed as *Incremental Learning with Alternating Strata* (ILAS). This mechanism splits the training dataset into several non-overlapping subsets of examples, and selects a different subset at each GA iteration. Thus, ILAS permits to reduce the training time of a single iteration of the GA since a lower number of examples need to be match with the new individuals in the evaluation process. Moreover, in [4], it was empirically shown that this technique allows for a better generalization.

3.3 SLAVE

SLAVE [10, 28] is an inductive learning algorithm based on a fuzzy-rule representation. The system follows an iterative rule learning scheme. In the following, the knowledge representation and the learning interaction are explained.

Knowledge Representation

SLAVE creates a set of individuals whose condition is represented in *conjunctive normal form*:

$$\textbf{IF } x_1 \text{ is } \widetilde{A_1^k} \wedge \ldots \wedge x_n \text{ is } \widetilde{A_n^k} \textbf{ THEN } c^k$$

where each input variable x_i is represented by a disjunction of linguistic terms $\widetilde{A_i^k} = \{A_{i1} \text{ or ... or } A_{in_i}\}$, and the consequent c^k is the class predicted by the rule. In our experiments, all the variables share the same semantics, which are defined by means of triangular-shaped fuzzy membership functions. The *matching degree* $\mu_{A^k}(e)$ of an example e with a classifier k is computed as follows. For each variable x_i, we compute the membership degree for each of its linguistic terms, and aggregate them by means of a T-conorm (disjunction). Then, the matching degree of the rule is determined by the T-norm (conjunction) of the matching degree of all the input variables.

In the inference process, the class of a new example is determined by the rule that maximizes the matching degree with this example. In case of having more than one rule with maximum matching degree, the class of the rule with maximum fitness is selected as the output.

Learning Interaction

SLAVE iteratively evolves a set of individuals following an iterative rule learning scheme [47]. This process is based on the iteration of the following steps: (a) learn one rule from the dataset, (b) penalize the data covered by the rule.

Figure 1 illustrates the learning scheme of SLAVE. Given a data set E and a specific class B, the system searches for the rule that describes this class more accurately. This process is performed by a steady-state genetic algorithm. The fitness function of the GA is determined by the train error and the generality of the rule. Then, this rule is aggregated to the fuzzy-rule set. If more rules are required to represent all the examples of class B, the examples covered by the current rules of class B are removed, and the GA is run again providing a new rule for class B. The same procedure is repeated until no more rules are required for class B. Then, the same algorithm is followed to learn rules for the other classes of the domain, resulting in a rule set that covers all the instances in the training data set.

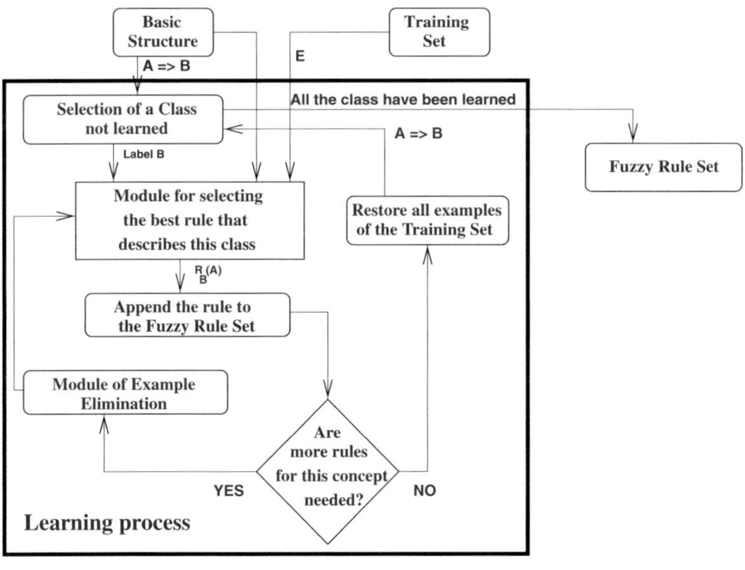

Fig. 1. Illustrative scheme of the learning process of SLAVE

3.4 Fuzzy AdaBoost

Fuzzy AdaBoost [14] is an evolutionary boosting algorithm that applies the AdaBoost algorithm [21] to learn a fuzzy-rule-based classifier. In the following, we introduce the knowledge representation and the learning procedure of Fuzzy AdaBoost.

Knowledge Representation

Fuzzy AdaBoost creates a set of *weak classifiers*, which take the following form:

$$\textbf{IF } x_1 \text{ is } A_1^k \wedge ... \wedge x_n \text{ is } A_n^k \textbf{ THEN } c_1^k \textbf{ WITH } s_1^k, \text{ ... }, c_p^k \textbf{ WITH } s_p^k$$

Each input variable x_i is represented by a *linguistic term* A_i^k. All variables share the same semantics, represented by triangular-shaped membership functions. The method permits the absence of a variable by not assigning any linguistic term to this variable. In the consequent, the rule maintains one weight s_j^k for each class j that indicates the soundness with witch the weak classifier predicts the class j.

The fuzzy inference is as follows. Given an input instance e, it is assigned to the class c that maximizes the following expression:

$$\text{arg max}_{k=1,\cdot,p} \bigvee_{j=1}^{N} A^j(e) \wedge s_k^j \tag{4}$$

where N is the number of classifiers in the population. $\vee$ and $\wedge$ are the t-norm and the t-conorm operators respectively.

Learning Interaction

Fuzzy AdaBoost iteratively invokes an algorithm that provides a *low quality classifier*, addressed as *weak hypothesis* in the boosting literature. Each example i in the training dataset has associated a weight w^i, and the search for promising classifiers focuses on the examples that have higher weights. Each time that a new weak hypothesis is added to the compound classifier, the examples in the training dataset are re-weighted. In that way, in next iterations, the search will be focused toward examples that are more difficult to learn. Moreover, at the end of each iteration, a voting strength α is assigned to each weak hypothesis, which depends on the confidence in the classification accuracy of that rule. In the following, these three steps briefly explained.

1. *Creation of weak classifiers* Fuzzy AdaBoost uses a integer-coded genetic algorithm [26] to evolve a population of the aforementioned weak classifiers. The best weak classifier generated by the GA is added to the compound classifier. The fitness of each classifier is computed as a combination of

two objectives. The first objective considers the generality of the rule, i.e., the proportion of examples that matches with high degree. The second objective computes the accuracy of the rule. Thus, the GA pressures toward highly general and accurate rules.

2. *Computation of the strength of a classifier* Fuzzy AdaBoost updates the strength α^j of a classifier according to the error of the rule generated in the previous generation j. That is, low errors result in high values of α^j; that means that the rule will have a strong influence in the inference of the class.
3. *Update of the weights of the training examples* After one iteration of the learning algorithm (and so, the generation of a new rule), the weights w^i of the examples in the training dataset are updated. Specifically, the weight of a correctly classified instance w_i is decreased according to the matching degree of the new rule with this instance and the strength α^j of the rule. On the other hand, the weight of incorrectly classified instances is increased, so that new rules will focus on the classification of these instances.

Class Inference of Test Examples

Given a new unknown example e, Fuzzy AdaBoost predicts the output of e as follows. Each rule j votes for each class it predicts according to (4), resulting in a vote $g^c(e)$ for a class c. Then, the votes for each class are added:

$$\text{vote}_k = \sum_{j:class_j=k} \alpha^j \cdot g^j(e) \tag{5}$$

The most voted class is returned as output.

3.5 Fuzzy LogitBoost

The LogitBoost algorithm [22] is a boosting method similar to AdaBoost that uses a greedy version of generalized backfitting [22] to build an additive model. It has been experimentally shown that LogitBoost outperforms AdaBoost, especially in multi-class problems. Due to these improvements, LogitBoost was extended to induce fuzzy classifiers, resulting in the so-called Fuzzy LogitBoost algorithm [36]. This learning technique inherits the main components from Fuzzy AdaBoost – such as the knowledge representation, the learning scheme, and the voting scheme – , but introduces the corrections proposed by LogitBoost. In the following, we detail the learning interaction, and refer the reader to Sect. 3.4 for details on the knowledge representation and class inference of new input examples.

Learning Interaction

Fuzzy LogitBoost follows a learning scheme similar to AdaBoost. The goal of the algorithm is to minimize the likelihood of the compound classifier. For details on the statistical formulation of the problem, the reader is referred to [22, 36, 42]. Instead, this section presents the final algorithm for Fuzzy LogitBoost.

Algorithm 3.1 shows the pseudocode for Fuzzy LogitBoost. The algorithm generates N weak classifiers, where N is a configuration parameter. The process to generate a new classifier is the following (see lines 4–22 of the algorithm). For each training example, the probability of the output class i, given the rule j and the example k is computed according to the function f_{ijk}, which represents the additive model. p_{ijk} is used to update the weigths w_{ijk}. Next, a genetic algorithm is applied to find a rule that fits accurately the training data, which is subject to the fitness function given in line 10. Then, the function values f_{ijk} are updated for each class i according to the best classifier found by the GA. Finally, the new rule is added to the compound classifier.

4 Experimentation

This section analyzes the competence of the five GBML systems in classification tasks. In pattern classification, we aim at obtaining accurate models which provide comprehensible explanations for human experts. For this purpose, we select a set of real-world problems and compare the performance and rule set interpretability of the GBML systems to a set of highly-competent and widely-used machine learning techniques. In the following, we first present the experimentation methodology and then compare the five GBML to the other learners.

Algorithm 3.1: Outline of the fuzzy logitboost algorithm

1 **Algorithm:** Logitboost

2 $f_{i0k} = 0$

3 **for** $j = 1, \cdots, N$ **do**

4   **for** $k = 1, \cdots, p$ **do**

5   **for** $i = 1, \cdots, n$ **do**

6             $p_{ijk} = \frac{e^{f_{ij-1k}}}{1+e^{f_{ij-1k}}}$

7             $w_{ijk} = p_{ijk}(1 - p_{ijk})$

8   **end**

9   **Find** rule antecedent A^j that minimizes $fitness(A^j)$, where:

10             $fitness(A^j) = \sum_i^n w_{ijk} \left(s^j \cdot A^j(x_i) - \frac{y_{ik} - p_{ijk}}{w_{ijk}} \right)^2$

11   **and**

12             $s^j = \frac{\sum_i (y_{ik} - p_{ijk}) A^j(x_i)}{\sum_i w_{ijk} [A^j(x_i)]^2}$

13   **for** $i = 1, \cdots, n$ **do**

14             $f_{ijk} = f_{ij-1k} + s^j \cdot A^j(x_i)$

15   **end**

16   **Aggregate** the new rule to the compound classifier

17   **end**

18 **end**

4.1 Methodology

Herein, we present (a) the real-world problems chosen for the experimentation, (b) the learning techniques included in the comparison, (c) the metrics used to evaluate the learners performance and models interpretability, and (d) the statistical analysis applied to evaluate the results.

Experimentation Problems

We selected a collection of 20 real-world problems with different characteristics (see Table 1) which may pose particular challenges to the different learning techniques. All these problems were obtained from the UCI repository [8], except for *tao*, which was chosen from a local repository [7].

Machine Learning Techniques Included in the Comparison

We compared the five GBML methods to C4.5, IBk, Naive Bayes, Part, SMO, and Zero-R. C4.5 [40] is a decision tree that enhances ID3 by introducing

Table 1. Properties of the datasets. The columns describe: the identifier of the dataset (Id.), the name of the dataset (dataset), the number of instances (#Inst), the total number of features (#Fea), the number of real features (#Re), the number of integer features (#In), the number of nominal features (#No), the number of classes (#Cl), and the proportion of instances with missing values (%MisInst)

Id.	Dataset	#Inst	#Fea	#Re	#In	#No	#Cl	%MisInst
ann	Annealing	898	38	6	0	32	5	0
aut	Automobile	205	25	15	0	10	6	22.4
bal	Balance	625	4	4	0	0	3	0
bpa	Bupa	345	6	6	0	0	2	0
cmc	Contraceptive method choice	1,473	9	2	0	7	3	0
col	Horse colic	368	22	7	0	15	2	98.1
gls	Glass	214	9	9	0	0	6	0
h-c	Heart-c	303	13	6	0	7	2	2.3
h-s	Heart-s	270	13	13	0	0	2	0
irs	Iris	150	4	4	0	0	3	0
pim	Pima	768	8	8	0	0	2	0
son	Sonar	208	60	60	0	0	2	0
tao	Tao	1,888	2	2	0	0	2	0
thy	Thyroid	215	5	5	0	0	3	0
veh	Vehicle	846	18	18	0	0	4	0
wbcd	Wisc. breast-cancer	699	9	0	9	0	2	2.3
wdbc	Wisc. diagnose breast-cancer	569	30	30	0	0	2	0
wne	Wine	178	13	13	0	0	3	0
wpbc	Wisc. prognostic breast-cancer	198	33	33	0	0	2	2
zoo	Zoo	101	17	0	1	16	7	0

methods to deal with continuous variables and missing values. IBk is an implementation of the nearest neighbor algorithm; it classifies a test instance with the majority class of its k nearest neighbors. Naive Bayes [33] is a probabilistic classifier that estimates the parameters of a Bayesian model. Part [19] is a learning architecture that combines the extraction of rules from partial decision trees and the separate-and-conquer rule learning technique to create a rule-based classifier without performing global optimization. SMO [39] is a widely-used implementation of *support vector machines* [46]. Zero-R is a very simple classifier that always predicts the majority class in the training dataset; this learner is included in the comparison to show a performance baseline. Table 2 summarizes the main characteristics of the learners.

All the non-fuzzy methods except for GAssist and UCS were run using WEKA [53]. For GAssist, we used the open source code provided in [3]. For UCS, we ran our own code. All the open source methods were configured with the parameters values recommended by default. Moreover, the models for SMO and IBk were selected as follows. For SMO, we ran the system with polynomial kernels of order 1, 3, and 10. Then, we ranked the results obtained with the three polynomial kernels and chose the model that maximized the average rank: SMO with polynomial kernels of order 3. Additionally, we also supply the results of SMO with Gaussian kernels. The same process was followed for IBk. We ran the experiments with $\mathrm{k} = \{1, 3, 5\}$, and chose the configuration that maximized the average rank: $\mathrm{k} = 5$. UCS was configured with the following parameters (see [6, 35] for notation details): $numIter = 100{,}000$, $N = 6{,}400$, $acc_0 = 0.99$, $\nu = 10$, $\{\theta_{GA}, \theta_{del}, \theta_{sub}\} = 50$, $\chi = 0.8$, $\mu = 0.04$, $\delta = 0.1$, $P_{\#} = 0.6$.

Fuzzy AdaBoost and Fuzzy LogitBoost were run using KEEL [2]. Default parameters were used, except for the maximum population size which was set to 50. The results of SLAVE were supplied by the authors. They used exactly the same datasets and validation methodology as the other methods in the comparison. SLAVE used a steady-state GA with population size $N = 100$, and probabilities of crossover $\chi = 1.0$ and mutation $\mu = 0.4$. The GA run was stopped after 500 iterations without improvement. In all these fuzzy methods, we used five linguistic terms per variable, defined by triangular-shaped membership functions.

Comparison Metrics

The data models built by each learner were evaluated in terms of performance and interpretability. We measured the performance of the method with the test accuracy, i.e., the proportion of correct classifications on previously unseen examples. To obtain reliable estimates of the test accuracy, we used a ten-fold cross validation procedure [45].

The comparison of the interpretability of the models is more complicated since the methods included in the analysis use different knowledge representations. For this reason, we first identify two groups of learners. The first group

Table 2. Summary of the main characteristics of the learners included in the comparison: C4.5, IBk, Naive Bayes (NB), Part, SMO, Zero-R (0-R), UCS, GAssist (GAt), SLAVE (SLV), and the two boosting algorithms (Bst), Fuzzy AdaBoost and Fuzzy LogitBoost

	Paradigm	Knowledge Rep. and Inference Method
C4.5	Decision-tree induction	Decision-tree. *Inference*: class given by the corresponding leaf.
IBk	Instance-based learning	No general model. *Inference:* class determined by the majority class of the k nearest neighboors.
NB	Statistical Modeling	Probabilities of a Bayesian model. *Inference:* the output is the class with maximum probability.
Part	Rule induction based on decision-tree induction and a separate-and-conquer approach	Ordered list of rules. Continuous variables represented by float-coded attributes. *Inference:* the output is the class of the first matching rule in the ordered list.
SMO	Neural networks (support vector machines)	Weights of the support vector machines. *Inference:* The class is determined by the decision function represented by the SVM.
0-R	Majority-class predictor	No knowledge representation. *Inference:* Majority class in the training dataset.
UCS	Michigan-style GBML	Population of intervalar rules with a fitness or strength value. *Inference:* The output is the most voted class (weighted by fitness) among the matching rules.
GAt	Pittsburgh-style GBML	Ordered list of intervalar rules (intervals obtained via discretization). Use of a default rule. *Inference:* the output is the class of the first matching rule in the ordered list.
SLV	Iterative Rule GBML	Population of linguistic fuzzy rules. *Inference:* class determined by the rule with maximum matching.
Bst	Statistical Learning Theory and GBML	Population of linguistic fuzzy rules with a strength per class. *Inference:* The output is the most voted class (weighted by the strength) among the matching classifiers.

consists of those methods that build models that can be hardly interpreted (e.g., models represented with weights) and learners that do not create any data model, i.e., *lazy learners*. The second group comprises those learners that build interpretable models (e.g. trees and rule sets). We excluded the first group of learners from the comparison, and focused on the models created by the second group of learners. As this group consists of learners with different types of knowledge representation, we provide some information of the sizes of the models to qualitatively compare them. For tree-based learners, we supply the number of leaves. For the rule-based systems, we provide the total number of rules evolved. Finally, we qualitatively compare these results considering the type of rule sets created and their size.

Statistical Analysis

We statistically analyzed the performance of each learner following the procedure pointed in [15]. As suggested by the author, we avoided to use any parametric statistical test since they require that the data satisfy several strong conditions. Instead, all the statistical analysis is based on non-parametric tests.

We first applied a multi-comparison statistical procedure to test whether all the learning algorithms performed the same on average. Specifically, we used the Friedman test [23, 24], the non-parametric equivalent to the analysis of variance test ANOVA [18]. If the Friedman test rejected that all the learners performed the same on average, post-hoc test were applied. Our first concern was to compare the performance of each GBML system with respect to the performance of the other learners. For this purpose, we applied the non-parametric Bonferroni–Dunn test [17]. The Bonferroni–Dunn test defines that one learner performs significantly differently to a control learner if the corresponding average rank differs by, at least, a critical difference CD computed as

$$CD = q_\alpha \sqrt{\frac{n_l(n_l+1)}{6n_{ds}}} \tag{6}$$

where n_l is the number of learners, n_{ds} is the number of datasets, and q_α is the critical value based on the Studentized range statistic [43]. We illustrate the results of this test by showing the group of learners that perform equivalently to the control learner.

The Bonferroni–Dunn test is said to be conservative, specially as the number of learners increases or the number of datasets decreases, so that it may not detect significant differences although they actually exist. Nonetheless, we use this test in the first step of our analysis since it permits to detect groups of learners that truly perform differently from other learners. We later apply pairwise comparisons to detect further significant differences between learners that belong to the same group. We used the non-parametric Wilcoxon signed-ranks test [49] for pairwise comparisons, and provide the approximative p-values computed as indicated in [43].

4.2 Results

Comparison of the Performance

Table 3 details the test accuracies obtained by each learner on the 20 real-world problems. The average performance of Fuzzy AdaBoost for the problems

Table 3. Comparison of the test performance of all the machine learning techniques: C4.5, IB5, Naive Bayes (NB), Part, SMO with polynomial kernels of order 3 (SMO_3), SMO with Gaussian kernels (SMO_r), Zero-R (0-R), GAssist (GAt), SLAVE (SLV), Fuzzy AdaBoost (ABst), and Fuzzy LogitBoost (LBst). The two last rows show the average rank of each learning algorithm (Rnk), and its position in the rank (Pos)

	C4.5	IB5	NB	Part	SMO_3	SMO_r	0-R	UCS	GAt	SLV	ABst	LBst
ann	98.9	97.3	86.3	98.6	99.3	91.9	76.2	99.0	97.9	96.8	–	76.2
aut	80.9	64.0	58.8	74.4	78.1	45.5	32.6	77.4	68.6	70.7	–	32.6
bal	77.4	88.2	90.6	82.9	91.2	88.3	45.5	77.3	79.6	72.0	85.5	88.3
bpa	62.3	58.8	56.0	67.6	60.0	58.0	58.0	67.5	62.2	60.0	65.3	64.5
cmc	52.6	46.5	50.6	50.0	48.7	42.7	42.7	50.3	53.6	46.1	49.5	51.1
col	85.3	81.5	78.2	84.5	75.6	82.4	63.1	96.3	94.3	82.9	63.1	63.1
gls	66.1	64.7	48.9	66.6	66.1	35.7	35.7	70.0	65.1	57.6	62.5	68.2
h-c	78.5	83.2	82.8	74.2	78.6	82.5	54.5	79.7	80.1	77.9	60.4	62.1
h-s	79.3	80.7	83.3	80.0	78.9	82.6	55.6	74.6	77.7	76.3	57.6	59.3
irs	94.0	96.0	96.0	94.0	92.7	93.3	33.3	95.4	96.2	93.3	95.5	95.3
pim	74.2	73.3	75.8	74.9	76.7	65.1	65.1	74.6	73.8	72.3	70.7	71.8
son	71.1	84.0	69.7	74.4	85.5	69.3	53.4	76.5	75.8	73.1	46.6	53.4
tao	95.9	97.1	81.0	94.3	84.2	83.6	49.9	87.0	91.6	82.1	91.5	91.7
thy	94.9	94.8	97.2	94.3	88.9	69.8	69.8	95.1	92.5	91.3	97.4	97.1
veh	71.1	68.9	46.3	73.4	83.3	41.7	25.4	71.4	67.0	66.5	30.8	37.2
wbcd	95.0	97.1	96.1	95.7	96.4	96.1	65.5	96.3	95.6	96.4	94.9	94.1
wdbc	94.4	96.8	93.1	94.5	97.6	92.9	63.1	96.0	94.2	91.7	37.3	62.7
wne	93.9	96.7	97.2	93.3	97.7	39.9	39.9	96.1	93.2	94.3	85.6	85.0
wpbc	71.6	78.9	69.5	70.0	81.3	73.0	73.0	69.4	72.3	75.7	23.6	76.4
zoo	92.8	90.5	94.5	93.8	97.8	76.0	41.9	96.8	94.0	96.5	41.9	41.9
Rnk	***5.25***	***4.68***	***5.88***	***5.13***	***4.23***	***8.18***	***11.08***	***4.4***	***5.45***	***7.13***	***8.95***	***7.68***
Pos	*5*	*3*	*7*	*4*	*1*	*10*	*12*	*2*	*6*	*8*	*11*	*9*

ann and *aut* is not provided since the system was not able to extract competent fuzzy rules from the two domains, leaving nearly all the feature space uncovered. We confirmed with the authors that this behavior could be due to the high number of nominal attributes that these two problems have. The last two rows of the table supply the average rank and the position of each algorithm in the ranking. The ranks were calculated as follows. For each dataset, we ranked the learning algorithms according to their performance; the learner with the highest accuracy held the first position, whilst the learner with the lowest accuracy held the last position of the ranking. If a group of learners had the same performance, we assigned the average rank of the group to each of the learners in the group.

The experimental results show the competitiveness of UCS and GAssist. UCS presents the second best average rank; it is only outperformed by SMO with polynomial kernels of order 3. GAssist holds the sixth position of the ranking, degrading the average rank obtained by SMO with polynomial kernels, UCS, IB5, Part, and C4.5. Finally, SLAVE, Fuzzy LogitBoost, and Fuzzy AdaBoost are in the position 8, 9, and 11 of the ranking.

We analyzed if these differences in the average ranks of the learners were statistically significant by means of the multi-comparison test of Friedman. The statistical test rejected the hypothesis that all the methods performed the same on average with $p = 7.84 \cdot 10^{-12}$. To evaluate the differences among them, we applied different statistical tests. First, we compared the five GBML systems to all the other learners by means of a Bonferroni–Dunn test at a significance level of 0.1. Figure 2 graphically represents the groups of learners that perform equivalently to (a) UCS, (b) GAssist (GAt), (c) SLAVE (SLV), (d) Fuzzy LogitBoost (LBst), and (e) Fuzzy AdaBoost (ABst). We did not include Zero-R in the analysis, since its results are only provided as a baseline.

The statistical analysis confirms the robustness of UCS and GAssist. UCS belongs to the group of the best learners, significantly outperforming SLAVE, Fuzzy LogitBoost, SMO with Gaussian kernels, and Fuzzy AdaBoost. The test also indicates that GAssist does not significantly degrade the performance of SMO with polynomial kernels, the best ranked method. Moreover, GAssist significantly outperforms SMO with Gaussian kernels and Fuzzy AdaBoost.

As the Bonferroni–Dunn test is said to be quite conservative [15], so that it may not detect all the significant differences, we complemented the statistical study by comparing each pair of learners. Table 4 shows the approximate p-values of the pairwise comparison according to a Wilcoxon signed-ranks test. The symbols $\oplus$ and $\ominus$ indicate that the method in the row significantly improves/degrades the performance obtained with the method in the column. Similarly, the symbols $+$ and $-$ are used to denote a non-significant improvement/degradation. Furthermore, Fig. 3 graphically illustrates the significant differences between methods. That is, each method is depicted in one vertex of the graph, and significant improvements (at $\alpha = 0.05$) of one learner with respect to another are plotted with a directed edge labeled with the corresponding p-value. To facilitate the visualization, the last three methods in

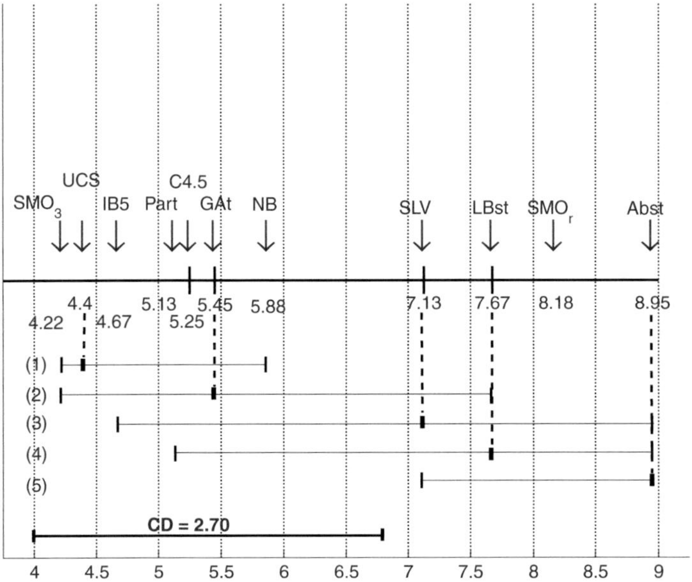

Fig. 2. Comparisons of one learner against the others with the Bonferroni–Dunn test at a significance level of 0.1. All the learners are compared to five different control groups: (1) UCS, (2) GAssist (GAt), (3) SLAVE (SLV), (4) Fuzzy LogitBoost (LBst), and (5) Fuzzy AdaBoost (ABst). The learners connected are those that perform equivalently to the control learner

Table 4. Pairwise comparisons of the learners by means of a Wilcoxon signed-ranks test. The above diagonal contains the approximate p-values. The below diagonal shows a symbol ⊕ / ⊖ if the method in the row significantly outperforms/degrades the method in the column at a significance level of .05 and +/=/− if there is no significant difference and performs better/equal/worse

	C4.5	IB5	NB	Part	SMO$_{p3}$	SMO$_{rbf}$	0-R	UCS	GAst	SLV	ABst	LBst
C4.5		.709	.263	.904	.421	.007	.000	.204	.941	.012	.001	.011
IB5	=		.053	.794	.412	.000	.000	.852	.455	.019	.002	.005
NB	−	−		.247	.067	.033	.000	.135	.108	.601	.003	.021
Part	+	−	+		.296	.006	.000	.232	.433	.014	.001	.006
SMO$_3$	+	+	+	+		.003	.000	.654	.296	.004	.003	.007
SMO$_r$	⊖	⊖	⊖	⊖	⊖		.001	.006	.003	.027	.167	.576
0-R	⊖	⊖	⊖	⊖	⊖	⊖		.000	.000	.000	.249	.001
UCS	+	=	+	+	−	⊕	⊕		.218	.002	.002	.007
GAt	−	−	+	−	−	⊕	⊕	−		.017	.002	.017
SLV	⊖	⊖	+	⊖	⊖	⊕	⊕	⊖	⊖		.023	.057
ABst	⊖	⊖	⊖	⊖	⊖	−	+	⊖	⊖	⊖		.004
LBst	⊖	⊖	⊖	⊖	⊖	−	⊕	⊖	⊖	−	⊕	

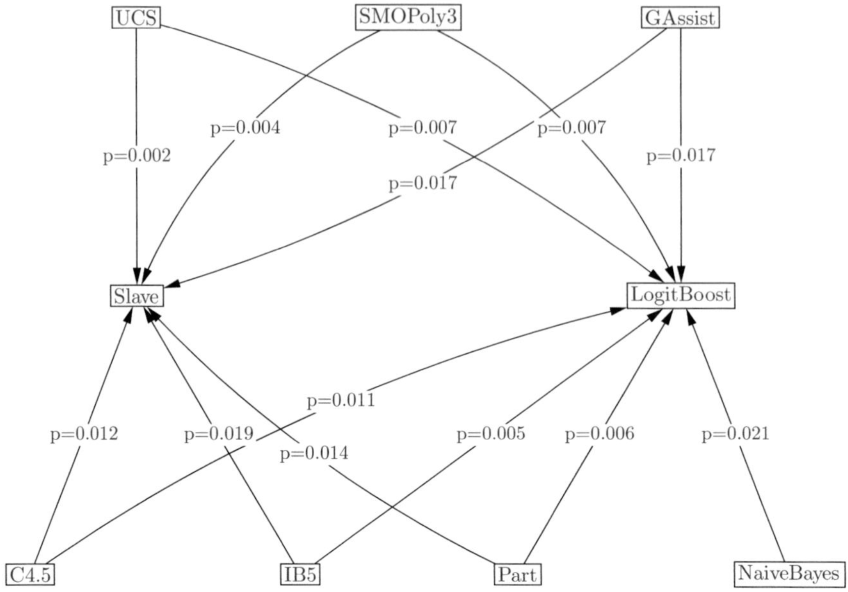

Fig. 3. Illustration of the significant differences of performance among methods. An edge $L_1 \stackrel{p_{value}}{\rightarrow} L_2$ indicates that the learner L_1 outperforms the learner L_2 with the corresponding p_{value}. To facilitate the visualization, SMO with Gaussian kernels, Fuzzy AdaBoost, and Zero-R are not included in the graph

the ranking are not included in the graph, i.e., SMO with Gaussian kernels, Fuzzy AdaBoost, and Zero-R. These three methods are outperformed by all the other methods.

The pairwise comparison confirms the conclusions derived from the Bonferroni–Dunn test, and finds further significant differences between pairs of learners. UCS, GAssist, SMO with polynomial kernels, Part, IB5, and C4.5 are the best methods in the comparison; any of them degrades the results obtained by other methods, and all them significantly outperform SLAVE, Fuzzy LogitBoost, SMO with Gaussian kernels, Fuzzy AdaBoost, and Zero-R. Thus, these results support that the two non-fuzzy GBML systems are, at least, as good as some of the most-used machine learning techniques.

The non-fuzzy GBML techniques present poorer results. Among them, the best method is SLAVE, with an average rank of 7.13. SLAVE is significantly outperformed by the six best methods of the analysis. Fuzzy LogitBoost is significantly outperformed by the same six methods and Naive Bayes. Fuzzy AdaBoost significantly degrades the results of all the other learners, except for SMO with Gaussian kernels, Fuzzy LogitBoost, and Zero-R. These results are not surprising. The linguistic fuzzy representation implies the discretization of the feature space, and the discretization points are fixed by the number of linguistic terms. In our experiments, we used only five linguistic terms per variable. Thus, in this scenario, a single rule may not have the required granularity to define accurately the class boundaries of a given complex domain, limiting the maximum accuracy that can be achieved. However, the linguistic fuzzy representation allows for a better readability of the rules, as discussed in the next section.

Comparison of the Interpretability

Figure 4 plots partial examples of the models created by each learner (except for IBk, which does not create any model) for the two-dimensional *tao* problem. The picture highlights the differences between knowledge representations, which make difficult the comparison among them. In our study, we provide some characteristics of the models to qualitatively evaluate their readability.

We first distinguish between two types of learners (a) lazy learners or learners with knowledge representation based on weights, and (b) rule-based or tree-based learners. As before, we exclude Zero-R from the comparison. The first group consists of IBk, SMO, and Naive Bayes. IBk is a lazy classifier that does not create any global model from the training dataset; to predict the output of a new input example, IBk searches for the k nearest neighbors and returns the majority class among them. SMO represents the knowledge by $\binom{n}{2}$ support vector machines (where n is the number of classes of the classification problem), each one consisting of a set of real-valued weights (see Fig. 4a). Naive Bayes builds models formed by a set of parameters which estimate the independent probabilities of a Bayesian model. Consequently, the knowledge created by these three methods is really hard to interpret. Without a further comparison among them, we state that these three methods provide the poorest models in terms of interpretability.

The second group comprises the tree-based learner C4.5, and the rule-based systems Part, UCS, GAssist, SLAVE, Fuzzy AdaBoost, and Fuzzy LogitBoost. C4.5 evolves a tree in which each node represents a decision over one variable (see Fig. 4b). Part creates a set of rules which are defined by a conjunction of conditions over their variables, and are interpreted as an ordered activation list (see Fig. 4c). The knowledge representation of GAssist, UCS, SLAVE, Fuzzy AdaBoost, and Fuzzy LogitBoost is detailed in Sect. 3, and examples are shown in Figs. 4d–g. It is worth noting the differences between these systems. The rule sets evolved by GAssist are interpreted as an ordered activation list, similarly to Part; besides, GAssist uses a default rule. On the other hand, in UCS, SLAVE, Fuzzy AdaBoost, and Fuzzy LogitBoost, rules are not ordered; they represent independent classifiers. Consequently, all the matching rules participate in the classification of new input instances (except for SLAVE, the inference process of which only considers the rule with maximum matching degree).

To evaluate the size of the models, we used the following metrics. For the tree-based system, we counted the number of decision leaves. For the rule-based systems, we used the number of rules. Note that these measures are not

```
-    1.000    * <0.229 0.875> * X]
-    0.298    * <0.708 0.437 > * X]
                     ...
```

(a) SMO

```
x <= -2.75
|  x <= -3.25: red (308.0)
|  x > -3.25
|  |  y <= 1.75: red (55.0)
|  |  y > 1.75
|  |  |  x <= -3: red (11.0/1.0)
|  |  |  x > -3
|  |  |  |    y < = 4.25: blue (6.0)
|  |  |  |    y > 4.25: red (4.0)
                    ...
```

(b) C4.5

if x $\leq$ -3.25 **then** red (308)
else if x $>$ 2.75 **then** blue (347/1)
else if y $\leq$ 0 **and** x $\geq$ -1 **then** red (192/1)
...

(c) Part

if x $>$ 2.72 **and** (y is [0.92,4.61] or y $>$ 5.07) **then** blue
else if (x is [-0.54, 0.54] or x $>$ 2.72) **and** y is [-4.28, -2.57] **then** blue
...
otherwise red

(d) GAssist

if x is [-6.00, -0.81] **and** y is [-6.00, 0.40] **then** red **with** acc = 1.00
if x is [2.84, 6.00] **and** y is [-5.26, 4.91] **then** blue **with** acc = 1.00
if x is [-6.00, -0.87] **and** y is [-6.00, 0.74] **then** red **with** acc = 1.00
...

(e) UCS

if x is {XS or S or M} **and** y is {XS or S or M} **then** red
if x is XS **then** red
if x is {VL or M or H or VH} **then** blue
...

(f) SLAVE

if x is L **and** y is L **then** blue **with** -5.42 **and** red **with** 0.0
if x is M **and** y is XS **then** blue **with** 2.21 **and** red **with** 0.0
if x is M **and** y is XL **then** blue **with** -2.25 **and** red **with** 0.0
...

(g) Boosting

Fig. 4. Examples of part of the models evolved by (**a**) SMO, (**b**) C4.5, (**c**) Part, (**d**) GAssist, (**e**) UCS, (**f**) SLAVE, and (**g**) Fuzzy AdaBoost and Fuzzy LogitBoost for the two-dimensional tao problem

Table 5. Average sizes of the models built by C4.5, Part, UCS, GAssist, SLAVE, Fuzzy AdaBoost (ABst), and Fuzzy LogitBoost (LBst)

	C4.5	Part	UCS	GAssist	SLAVE	ABst	LBst
ann	38	15	4,410	5	8	50	50
aut	44	21	4,064	7	17	50	50
bal	45	37	1,712	8	22	50	50
bpa	25	9	2,603	6	6	50	50
cmc	162	168	3,175	15	49	50	50
col	5	9	3,446	5	7	50	50
gls	24	15	3,013	5	15	50	50
h-c	29	21	2,893	6	6	50	50
h-s	17	18	3,499	5	7	50	50
irs	5	4	634	3	3	50	50
pim	19	7	3,225	7	13	50	50
son	14	8	5,999	5	9	50	50
tao	36	17	609	6	3	50	50
thy	8	4	1,283	4	5	50	50
veh	69	32	4,601	7	26	50	50
wbcd	12	10	1,799	3	5	50	50
wdbc	11	7	5,079	4	5	50	50
wne	5	5	3,413	3	4	50	50
wpbc	12	7	5,078	4	10	50	50
zoo	11	8	1,244	6	7	50	50

directly comparable due to the differences in the knowledge representations. However, we use these metrics to make a qualitative analysis.

Table 5 shows the model sizes of the rule-based and tree-based systems. These results show that:

- GAssist evolves rule sets that are significantly smaller than the rule sets created by all the other methods according to a Wilcoxon signed-rank test (at $\alpha = 0.05$). Thus, in terms of population size, GAssist is the best method in the comparison.
- SLAVE builds the second smallest rule sets, being only improved by GAssist. However, note that the types of rules evolved by SLAVE are much easier to read for two main reasons:
  1. GAssist uses an intervalar representation, where the intervals are obtained by applying a discretization technique over the input space. On the other hand, SLAVE uses a linguistic fuzzy representation. That is, variables are represented by linguistic terms; so, the rules can be easily read by human experts.
  2. GAssist uses an ordered activation list; that is, a rule is used to infer the class of a new input instance only if all the previous rules in the list do not match with this instance. Thus, the context of the rules (i.e., the conditions of the previous rules in the activation list) has to

be considered to read the whole rule set. Oppositely, in SLAVE, rules are independent classifiers. So, human experts can read each rule individually. This supposes one of the main advantages of the knowledge representation of SLAVE with respect to the representation of GAssist.
3. GAssist uses a default rule, which reduces considerably the rule set size, while SLAVE does not. The rule set sizes of SLAVE could be further reduced by including a default rule.

For these three reasons, we consider that the models created by SLAVE are more interpretable than those evolved by GAssist, even though the rule sets evolved by SLAVE are slightly bigger than those created by GAssist.

- Fuzzy AdaBoost and Fuzzy LogitBoost create rule sets of moderate size. In fact, the size of these rule sets is determined by a configuration parameter. In our experiments, we set the maximum population size to 50 since it maximized the average performance rank of the two learners. Smaller population sizes could be set for few specific problems without loss of accuracy. However, in our analysis, we are interested in robust methods that do not highly depend on the configuration parameters. For this reason, we used the same parameters in all runs, and did not search for the best configuration of each system for each particular dataset.

  This identifies a disadvantage of the two statistical methods with respect to the other learners. While UCS, GAssist, and SLAVE evolve a different number of rules depending on the intrinsic complexities of the domain, Fuzzy AdaBoost and Fuzzy LogitBoost need to know beforehand the number of rules to be created. Several techniques could be applied to overcome this drawback. For example, new approaches could be designed to remove the new rules that do not improve the accuracy of the compound classifier.
- UCS evolves the largest rule sets among all the methods in the comparison. Thus, even using a rule-based representation, the high number of rules may hinder the interpretability of the models evolved. Some reduction techniques have been proposed to remove non-useful rules from the final populations in XCS [16,25,52]. In further work, these reduction techniques will be adapted to UCS to try to improve the interpretability of the models.

The whole study performed through this section highlights that *genetic-based machine learning* is one of the best alternatives for facing challenging data mining and classification problems. The analysis also provides guidelines on which system should be used depending on the requirements of the problem. If the accuracy of the classification model is crucial, UCS appears to be the best approach to face a new problem. If the results also need to be readable, our recommendation is to use GAssist, since it offers highly interpretable models that are nearly as accurate as those created by UCS. If the readability prevails over the performance, SLAVE turns out to be the best choice as it creates a very low number of easily-readable independent linguistic fuzzy rules.

5 Summary and Conclusions

In this paper, we studied the suitability of *genetic-based machine learning* for pattern classification. For this purpose, we selected five GBML approaches that represent different styles and knowledge representations; two non-fuzzy GBML systems: UCS, GAssist; and three fuzzy GBML methods: SLAVE, Fuzzy AdaBoost, and Fuzzy LogitBoost. These learners were compared to six highly-used machine learning techniques on a set of 20 real-world problems. The results were analyzed by means of different statistical procedures.

The analysis showed the competence of GBML for classification tasks, and also pointed some recommendations on which GBML system use depending on the requirements of the user. UCS showed to be one of the best learners in terms of performance. It presented the second best performance, only being outperformed by the support vector machine SMO with polynomial kernels of order 3. GAssist resulted in a slightly inferior average performance, but evolved much more readable rule sets. SLAVE evolved the most interpretable models, but their performance was significantly inferior than the performance of the models created by UCS and GAssist.

Another observation drawn from the comparison is that there is not any method that outperforms another learner in all the problems. This indicates that the intrinsic complexities of the classification domains may have different effects depending on the learner. Our proposal as further work is to include measures to evaluate the complexity of the domains in the comparison, with the aim of pointing the complexities that affect the performance of the different GBML systems.

Acknowledgements

The authors thank the support of *Enginyeria i Arquitectura La Salle*, Ramon Llull University, as well as the support of *Ministerio de Educación y Ciencia* under project TIN2005-08386-C05-04, and *Generalitat de Catalunya* under Grants 2005FI-00252 and 2005SGR-00302.

References

1. J. Aguilar-Ruiz, J. Riquelme, and M. Toro. Evolutionary Learning of Hierarchical Decision Rules. *IEEE Transactions on Systems, Man, and Cybernetics Part B*, 33(2):324–331, 2003.
2. J. Alcalá-Fdez, M.J. del Jesus, J.M. Garrell, F. Herrera, C. Herbás, and L. Sánchez. Proyecto KEEL: Desarrollo de una herramienta para el análisis e implementación de algoritmos de extracción de conocimiento evolutivos. In J.S. Aguilar R. Giráldez, J.C. Riquelme, editor, *Tendencias de la Minería de Datos en España, Red Española de Minería de Datos y Aprendizage*, pages 413–424, 2004.

3. J. Bacardit. *GAssist Source Code: http://www.asap.cs.nott.ac.uk/jqb/PSP/GAssist-Java.tar.gz.*
4. J. Bacardit. *Pittsburgh Genetic-Based Machine Learning in the Data Mining Era: Representations, generalization and run-Time.* PhD thesis, Ramon Llull University, Barcelona, 2004.
5. T. Bäck. *Evolutionary Algorithms in Theory and Practice: Evolution Strategies, Evolutionary Programming, Genetic Algorithms.* Oxford University Press, Oxford, 1996.
6. E. Bernadó-Mansilla and J.M. Garrell. Accuracy-Based Learning Classifier Systems: Models, Analysis and Applications to Classification Tasks. *Evolutionary Computation*, 11(3):209–238, 2003.
7. E. Bernadó-Mansilla, X. Llorà, and J.M. Garrell. XCS and GALE: A Comparative Study of Two Learning Classifier Systems on Data Mining. In *Advances in Learning Classifier Systems*, volume 2321 of *LNAI*, pages 115–132. Springer, Berlin Heidelberg New York, 2002.
8. C.L Blake and C.J. Merz. *UCI Repository of ML Databases: http://www.ics.uc.edu/mlearn/MLRepository.html.* University of California, 1998.
9. P. Bonelli and A. Parodi. An efficient classifier system and its experimental comparison with two representative learning methods on three medical domains. In *4th International Conference on Genetic Algorithms*, pages 288–295, 1991.
10. L. Castillo, A. González, and R. Pérez. Including a simplicity criterion in the selection of the best rule in a genetic fuzzy learning algorithm. *Fuzzy Sets and Systems*, 120:309–321, 2001.
11. O. Cordón, F. Herrera, F. Hoffmann, and L. Magdalena. *Genetic Fuzzy Systems: Evolutionary Tuning and Learning of Fuzzy Knowledge Bases*, volume 19 of *Advances in Fuzzy Systems–Aplications and Theory.* World Scientific, Singapore, 2001.
12. K.A. de Jong and W. Spears. Learning Concept Classification Rules Using Genetic Algorithms. In *Proceedings of the International Joint Conference on Artificial Intelligence*, pages 651–656. Sydney, Australia, 1991.
13. K.A. de Jong, W.M. Spears, and D.F. Gordon. Using Genetic Algorithms for Concept Learning. *Genetic Algorithms for Machine Learning, A Special Issue of Machine Learning*, 13, 2–3, pages 161–188, 1993.
14. M.J. del Jesús, F. Hoffmann, L.J. Navascués, and L. Sánchez. Induction of fuzzy-rule-based classifiers with evolutionary boosting algorithms. *IEEE Transactions on Fuzzy Systems*, 12(3):296–308, 2004.
15. J. Demšar. Statistical Comparisons of Classifiers over Multiple Data Sets. *Journal of Machine Learning Research*, 7:1–30, 2006.
16. P.W. Dixon, D.W. Corne, and M.J. Oates. *A Ruleset Reduction Algorithm for the XCSI Learning Classifier System*, volume 2661/2003 of *Lecture Notes in Computer Science*, pages 20–29. Springer, Berlin Heidelberg New York, 2004.
17. O.J. Dunn. Multiple Comparisons among Means. *Journal of the American Statistical Association*, 56:52–64, 1961.
18. R.A. Fisher. *Statistical Methods and Scientific Inference*, 2nd edn. Hafner Publishing Company, New York, 1959.
19. E. Frank and I.H. Witten. Generating accurate rule sets without global optimization. In *Proceedings of the 15th International Conference on Machine Learning*, pages 144–151. Morgan Kaufmann, San Francisco, 1998.
20. A. Freitas. *Data Mining and Knowledge Discovery with Evolutionary Algorithms.* Springer, Berlin Heidelberg New York, 2002.

21. Y. Freund and R.E. Schapire. Experiments with a New Boosting Algorithm. In *International Conference on Machine Learning*, pages 148–156, 1996.
22. J. Friedman, T. Hastie, and R. Tibshirani. Additive logistic regression: a statistical view of boosting. *Annals of Statistics*, 32(2):337–374, 2000.
23. M. Friedman. The Use of Ranks to Avoid the Assumption of Normality Implicit in the Analysis of Variance. *Journal of the American Statistical Association*, 32:675–701, 1937.
24. M. Friedman. A Comparison of Alternative Tests of Significance for the Problem of m Rankings. *Annals of Mathematical Statistics*, 11:86–92, 1940.
25. C. Fu and L. Davis. A modified classifier system compaction algorithm. In *GECCO'02: Proceedings of the Genetic and Evolutionary Computation Conference*, pages 920–925. Morgan Kaufmann, San Francisco, 2002.
26. D.E. Goldberg. *Genetic Algorithms in Search, Optimization & Machine Learning*, 1st edn. Addison Wesley, Reading, 1989.
27. D.E. Goldberg. *The Design of Innovation: Lessons from and for Competent Genetic Algorithms*, 1st edn. Kluwer, Boston, 2002.
28. A. Gónzalez and R. Pérez. Completeness and Consistency Conditions for Learning Fuzzy Rules. *Fuzzy Sets and Systems*, 96:37–51, 1998.
29. A. Gónzalez and R. Pérez. SLAVE: A Genetic Learning System Based on an Iterative Approach. *IEEE Transactions on Fuzzy Systems*, 7(2):176–191, 1999.
30. J.H. Holland. *Adaptation in Natural and Artificial Systems.* The University of Michigan Press, Michigan, 1975.
31. J.H Holland. Escaping Brittleness: The possibilities of General-Purpose Learning Algorithms Applied to Parallel Rule-Based Systems. In Michalski Mitchell and Carbonell, editors, *Machine Learning, an artificial intelligence approach*, volume II of *Lecture Notes in Artificial Intelligence*, pages 593–623. Morgan Kaufmann, San Francisco, 1986.
32. C.Z. Janikow. A Knowledge-Intensive Genetic Algorithm for Supervised Learning. *Machine Learning*, 13(2–3):189–228, 1993.
33. G.H. John and P. Langley. Estimating Continuous Distributions in Bayesian Classifiers. In *11th Conference on Uncertainty in Artificial Intelligence*, pages 338–345. Morgan Kaufmann, San Francisco, 1995.
34. Z. Liu, A. Liu, C. Wang, and Z. Niu. Evolving neural network using real coded genetic algorithm (GA) for multispectral image classification. *Future Generation Computer Systems*, 20(7):1119–1129, 2004.
35. A. Orriols-Puig and E. Bernadó-Mansilla. A Further Look at UCS Classifier System. In *GECCO'06: Genetic and Evolutionary Computation Conference Workshop Program*, ACM Press, Seattle, 08–12 July 2006.
36. J. Otero and L. Sánchez. Induction of descriptive fuzzy classifiers with the logitboost algorithm. *Soft Computing*, 10(9):825–835, 2006.
37. M. Pelikan. *Hierarchical Bayesian Optimization Algorithm: Toward a New Generation of Evolutionary Algorithms*, volume 170 of *Studies in Computational Intelligence.* Springer, Berlin Heidelberg New York, 2005.
38. M. Pelikan, K. Sastry, and E. Cantú-Paz. *Scalable Optimization via Probabilistic Modeling*, volume 33 of *Studies in Computational Intelligence.* Springer, Berlin Heidelberg New York, 2006.
39. J. Platt. Fast Training of Support Vector Machines using Sequential Minimal Opt. In *Advances in Kernel Methods - Support Vector Lear.* MIT Press, 1998.
40. J.R. Quinlan. *C4.5: Programs for Machine Learning.* Morgan Kaufmann, San Mateo, California, 1995.

41. J. Rissanen. Modeling by shortest data description. *Automatica*, vol. 14:465–471, 1978.
42. R.E. Schapire and Y. Singer. Improved Boosting Algorithms using Confidence-Rated Predictions. *Machine Learning*, 37(3):297–336, 1999.
43. D.J. Sheskin. *Handbook of Parametric and Nonparametric Statistical Procedures.* Chapman & Hall, Boca Raton, 2000.
44. R.S. Sutton and A.G. Barto. *Reinforcement learning: An introduction.* MIT, Cambridge, 1998.
45. T.G. Dietterich. Approximate Statistical Tests for Comparing Supervised Classification Learning Algorithms. *Neural Computation*, 10(7):1895–1924, 1998.
46. V. Vapnik. *The Nature of Statistical Learning Theory.* Springer, Berlin Heidelberg New York, 1995.
47. G. Venturini. SIA: A Supervised Inductive Algorithm with Genetic Search for Learning Attributes Based Concepts. In P. B. Brazdil, editor, *Machine Learning: ECML-93 - Proceedings of the European Conference on Machine Learning*, pages 280–296. Springer, Berlin Heidelberg New York, 1993.
48. D. Wierstra, F.J. Gómez, and J. Schmidhuber. Modeling Systems with Internal State Using Evolino. In *GECCO'05: Proceedings of the 2005 conference on Genetic and evolutionary computation*, pages 1795–1802. ACM Press, New York, 2005.
49. F. Wilcoxon. Individual Comparisons by Ranking Methods. *Biometrics*, 1:80–83, 1945.
50. S.W. Wilson. Classifier Fitness Based on Accuracy. *Evolutionary Computation*, 3(2):149–175, 1995.
51. S.W. Wilson. Generalization in the XCS Classifier System. In *3rd Annual Conference on Genetic Programming*, pages 665–674. Morgan Kaufmann, San Francisco, 1998.
52. S.W. Wilson. Compact Rulesets from XCSI. In *Advances in Learning Classifier Systems, 4th International Workshop*, volume 2321 of *Lecture Notes in Artificial Intelligence*, pages 197–210. Springer, Berlin Heidelberg New York, 2002.
53. I.H Witten and E. Frank. *Data Mining: Practical Machine Learning Tools and Techniques*, 2nd edn. Morgan Kaufmann, San Francisco, 2005.